Mechanisms and Machine Science

Volume 92

This book series establishes a well-defined forum for monographs, edited Books, and proceedings on mechanical engineering with particular emphasis on MMS (Mechanism and Machine Science). The final goal is the publication of research that shows the development of mechanical engineering and particularly MMS in all technical aspects, even in very recent assessments. Published works share an approach by which technical details and formulation are discussed, and discuss modern formalisms with the aim to circulate research and technical achievements for use in professional, research, academic, and teaching activities.

This technical approach is an essential characteristic of the series. By discussing technical details and formulations in terms of modern formalisms, the possibility is created not only to show technical developments but also to explain achievements for technical teaching and research activity today and for the future.

The book series is intended to collect technical views on developments of the broad field of MMS in a unique frame that can be seen in its totality as an Encyclopaedia of MMS but with the additional purpose of archiving and teaching MMS achievements. Therefore, the book series will be of use not only for researchers and teachers in Mechanical Engineering but also for professionals and students for their formation and future work.

The series is promoted under the auspices of International Federation for the Promotion of Mechanism and Machine Science (IFToMM).

Prospective authors and editors can contact Mr. Pierpaolo Riva (publishing editor, Springer) at: pierpaolo.riva@springer.com

Indexed by SCOPUS and Google Scholar.

More information about this series at http://www.springer.com/series/8779

Korganbay Sagnayevich Sholanov

Parallel Manipulators of Robots

Theory and Applications

IFToMM

Springer

Korganbay Sagnayevich Sholanov
Department of Automation and Technology
Karaganda State Technical University
Karaganda, Kazakhstan

ISSN 2211-0984 ISSN 2211-0992 (electronic)
Mechanisms and Machine Science
ISBN 978-3-030-56072-0 ISBN 978-3-030-56073-7 (eBook)
https://doi.org/10.1007/978-3-030-56073-7

This Springer imprint is published by the registered company Springer Nature Switzerland AG
The registered company address is: Gewerbestrasse 11, 6330 Cham, Switzerland

Introduction

Robot is a technical system that is capable of some independent mechanical work and having exterior or functional similarity with a human or some other living beings. If a robot is compared to a human, it is revealed that functions of human skeleton in robots are implemented by a mechanical system in the form of manipulators. Applied in the robot-controlled actuators, microprocessor systems, information, and sensor systems belong to embedded systems. These embedded systems of robots may be applied both separately and in aggregate, also in other facilities with automated control system. In this regard in the book, the main place is taken by characteristic for robots' executive mechanisms—manipulators that are distinguished by their diversity. Robot manipulators are multi-degree-of-freedom mechanisms. Kinematic chain of manipulator can be open, closed with one or several loops, or can include both open and closed loops. Recently, parallel manipulators have come into use; these manipulators consist of one or several closed loops [1–3]. This book deals with the issues of the design, kinematics, and dynamics of robot parallel manipulators. It also considers possible parallel fields of application. Today, multiloop manipulators with various architectures are widely used in the industry in the function of technological equipment; this allowed the creation of robotic tools with improved dynamic and technological capabilities. Multiloop manipulators are used in special-purpose controlled machines that are applied in various spheres of life activity, from sports and entertainment to military and space sphere. Field of possible application of robots is constantly extending; specified in this paper possible fields of application demonstrate the probable extension of field of application of robots with parallel manipulators.

The book consists of four parts.

The first part is devoted to the synthesis of the architecture of one and multicircuit manipulators. Analysis of studies of parallel robot architecture synthesis shows that available diversity of robot architectures is a set of solutions for individual problems and they do not have any systematic tendency. This part offers a systematic (reasoned) approach to the synthesis of parallel manipulators or multi-degree-of-freedom mechanisms of robots, and it also systematizes and divides them into one and multicircuit manipulators. In this regard, the paper considers separately the theory of robot one and multicircuit parallel manipulators' synthesis. As a result, a class of one-circuit parallel manipulator (OPM) and multicircuit parallel manipulator (MPM) were obtained.

The second part in detail describes the theory of analysis of robot parallel manipulator design. Herein the method based on the use of zero parameters of coordinate system change from kinematics is offered to determine the number of dependent links in the analysis of closed mechanisms' and manipulators' design. The zero-parameter method is useful for practical application; it allows for establishing actual degree-of-freedom of any closed mechanism or manipulator including those having versatility, based on the known correlations. The part describes a method used in the kinematics for choosing the coordinate systems related to the links, and the sequence of coordinate system transformation and identification of the transformation parameters for each kinematic pair.

The kinematic parameters for coordinate systems are analyzed, and it is justified that the zero parameters characterize the peculiarities in junction configuration. At the same time, the zero parameters can be determined indirectly without calculating certain values of other transformation parameters.

It is proved that the number of transformation parameters amounting to zero affects the rank of coefficients matrix of linear equations system obtained based on the dynamic general equation.

At the same time, it is established that the number of the dependent links is equal to the least number of zero parameters (coefficients amounting to zero) in each of the system's equations. To in practice determine the number of dependent links, it is offered to form a table for each closed contour with lines corresponding to the kinematic pairs, and with columns containing six transformation parameters for a corresponding kinematic pair. The table includes transformation parameters for fictitious kinematic pair formed between fixed members. At the same time, as it is stated above, the transformation parameters are not calculated, and attention is paid only to parameters that are equal to zero. As a result, the number of dependent links is determined as the least number of zero parameters when comparing all lines of the table formed for the closed kinematic chain. With the known number of dependent links, the degree-of-freedom of one and multicircuit closed mechanisms can be determined by the known dependences. As an example, zero-parameter method is used for analyzing the Bennett mechanism topology. At the same time, actual degree-of-freedom is obtained for the Bennett mechanism, which is not determined by any of the known formulas. Thus, this confirms the universality and correctness of the offered method that is based on determining number of dependent links using zero parameters of transformation. The example of eliminating the dependent links in the Bennett mechanism demonstrates the possibility for using zero-parameter method for the synthesis of mechanisms without dependent links.

The third part describes kinematics and dynamics of one and multicircuit parallel manipulators separately. The first section of the part deals with the method of analytical method of describing position of links of any mechanisms and manipulators, as well as obtaining of more simple final matrix expression (with smaller number of terms). It is proposed to transform the coordinate system via connecting straight line using eight values as transformation parameters. With such transformation, all geometric and kinematic parameters of links and KP number are among the transformation parameters. Redundant parameter method expands the field of application of matrix method for solving kinematic problem regarding the position of parallel manipulator links. This part considering the architecture separately performs the kinematic and dynamic analysis of OPM and MPM.

The fourth part consists of materials related to the use of one and multicircuit manipulators of robots. With the method of OPM synthesis described in Part I, over 20 types of OPM can be synthesized. One of the OPM types in the form of six-axis table is produced as a demonstration model (KazNTU robot). This model demonstrates that it is possible to create six-axis metal-working machinery for processing details with complex configuration. With just slight modifications, KazNTU robot can be used as a mixer for multicomponent mixtures. As it is stated earlier, architecture of the multiloop manipulators is often formed by intuition. This paper offers a new type of multiloop manipulator having a whole range of advantages if compared to the existing MPM. This part gives examples of possible use of the SHOLKOR MPM. At the same time for better credibility, each example of use is justified by the operative demonstration model. The paper provides examples of use of SHOLKOR MPM in medicine, seismic protection, and renewable energy.

Contents

Part I
Synthesis of Architecture of Robot One-Loop and Multiloop Manipulators

Chapter 1
Synthesis of Robot One-Loop Multi-Degree-of-Freedom Manipulators

Abstract Available diverse parallel manipulator design schemes are mainly derived by intuitive methods based on engineering practice. This chapter proposes a systematic (reasoned) approach to the synthesis of parallel manipulators, which are divided into classes of one-loop parallel manipulators (OPM) and multiloop parallel manipulators (MPM). Synthesis of OPM is based on systematization of manipulators with an open kinematic chain, allocation of core functional groups (CFG) and interconnection of CFG and connecting link. Using this method, it is possible to synthesize 20 schemes of OPM design.

Keywords Parallel manipulators · One-loop · Multiloop · Core functional groups

1.1 Systematization of Robot Manipulators Design

Available today variety of design schemes of robot manipulators is explained by both difference in the robot purpose, variety of technical, operational and other requirements, and difference (in most cases intuitive) in approaches to their design analysis and synthesis [4–12]. That is why, some systematization of manipulators design schemes is required.

Robot mechanisms design can be systematized according to their functional capabilities if we assume that any robot is based on core functional groups (CFG) [13–15]. Such a representation allows comparing the CFG by their functional capabilities and can be used in the design analysis and synthesis, as well as in kinematic and dynamic analysis.

Systematization of the robot mechanism design by its functional capabilities reasons from the following requirements: some point $\boldsymbol{M}$ of the output link (gripper) in any position must have such mobility that ensures transfer of this point in the space from this position to any following one. At the same time the manipulator that satisfies these requirements is deemed to be made of solid nondeformable bodies forming only rotational and translational pairs. The specified manipulators have zero maneuverability. Kinematic chain of such manipulators is called core functional group.

K. S. Sholanov, *Parallel Manipulators of Robots*, Mechanisms and Machine Science 92,
https://doi.org/10.1007/978-3-030-56073-7_1

Thus, core functional group is a kinematic chain made of three links and three one-DOF kinematic pairs ensuring any set movements of the gripper point having zero maneuverability when using principle of solidification and zero-DOFof external connections. Comparison of domestic and foreign robot mechanisms design shows that their architecture is based on certain CFG. Different options of robot mechanism architecture are combinations of CFG or contain several additional links and kinematic pairs connected to the base CFG to ensure the required functional capability. Systematization of the CFG design shows that the set movement of the gripper point can be obtained only in case of some combination, certain sequence and orientation of translational pairs, their guides, and rotational pairs, as well as their hinge axes; all these can be formed as certain rules. CFG design schemes and rules for building each of them are given in Table 1.1. At the same time the main ones are CFG-1,-2,-3 made of combinations of kinematic pairs TTT, RTT, RRT, for which direct and reverse problems of kinematic are solved unambiguously.

1.2 Synthesis of Robot One-Loop Manipulators

During the synthesis of the robot architecture [15, 16], we will reason from that the output link of the robot could move as a controlled free body. The mechanics knows the Charles' theorem for free motion, according to which free motion of a body is considered as forward movement with a pole and rotation around some axis passing through this pole. This representation is often used in robotic engineering. For example, motions performed by the actuators are divided into transportation motions of the gripper center and orienting motions around the axis that pass through the gripper center. This paper mainly uses other representation of free motion of the robot working body. More specifically, in order to move working body as a free body, it is necessary to first move some axis of the working body (at the same time the axis must have degree of freedom of $W = 5$), and then rotate the body around this axis (rotation motion with one degree of freedom of $W = 1$). As a result the robot working body will have capability of moving as a free body with six degrees of freedom. This free movement analysis principle underlies the offered method of synthesis of parallel manipulators with one-loops (OPM) with DOF $W = 6$ [13–15]. CFG are used for the synthesis of design schemes. The problem of the synthesis of the design scheme of a multi-DOF mechanism is solved as the problem of developing design scheme for the robot mechanism, in which the working body (gripper) has six degrees of freedom and can rotate around the gripper axis of rotation. Design of the mechanism intended for performing these movements can be functionally broken into three components: into two mechanisms of transferring two points of the gripper axis moved and into the mechanism of rotation around this axis. According to its functional requirements, transfer mechanisms must ensure transfer in the operational scope of the set point with three degrees of freedom. Thus mechanism must have at least three degrees of freedom. There is a lot of mechanisms with kinematic chains, which satisfy this requirement. However, if kinematic chains consist of rotational

Table 1.1 Design schemes and CFG formation rules

No CFG	Design schemes CFG	CFG design rules
1	2	3
1 T⊥T⊥T	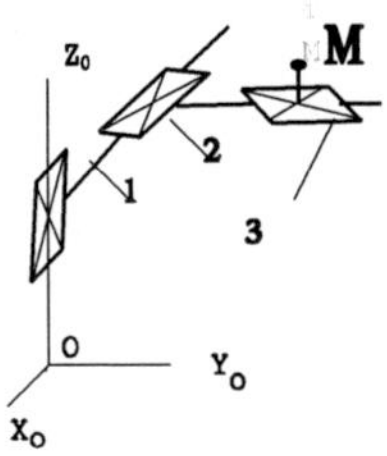	Velocity vectors in translational movement—noncomplanar vectors
2 R‖T⊥T	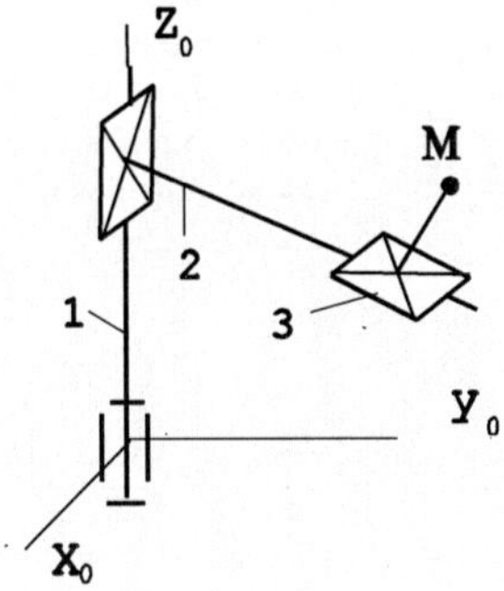	One of the velocity vectors in translational movement collinear to the rotational angular velocity vector, and the other is perpendicular to it
3 R⊥R⊥T	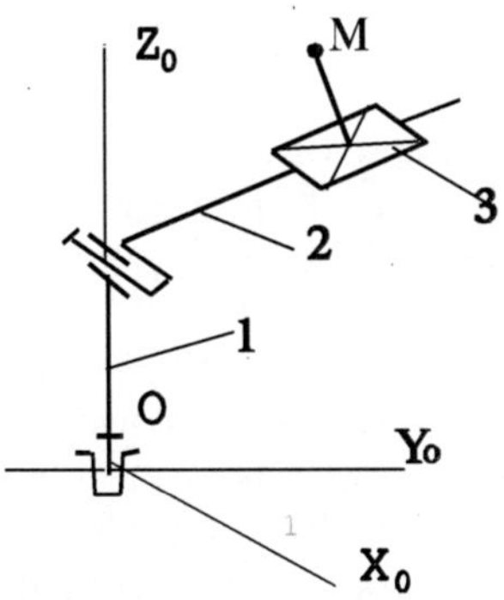	Velocity vector in translational movement forms noncomplanar system of vectors with non-collinear angle velocity vectors
4 T‖R‖R	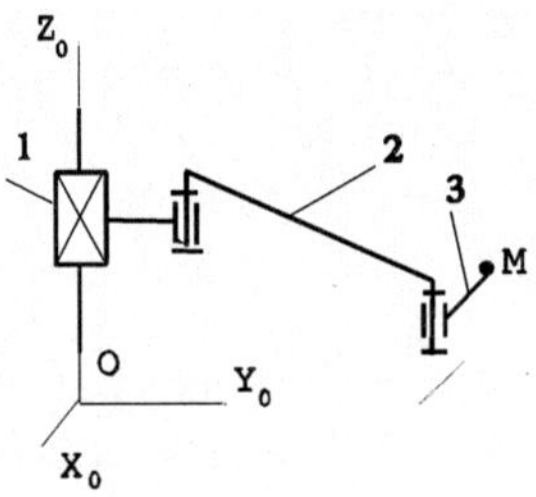	Angular velocity vectors and velocity vectors in translational movement—collinear vectors

(continued)

Table 1.1 (continued)

No CFG	Design schemes CFG	CFG design rules
5 R⊥R‖R	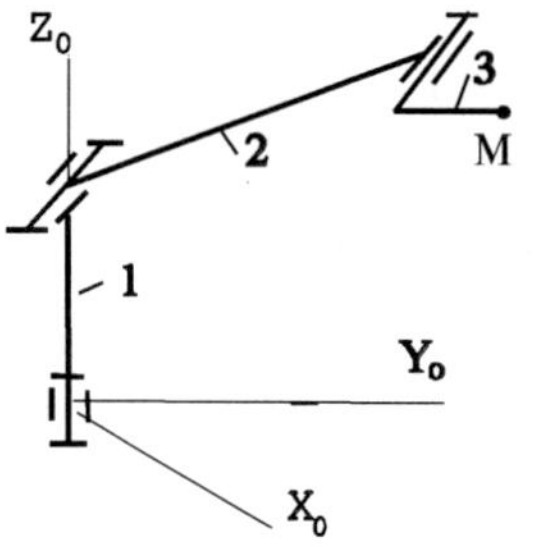	All angular velocity vectors—noncomplanar vectors, while two angular velocity vectors are collinear
6 T‖R⊥T	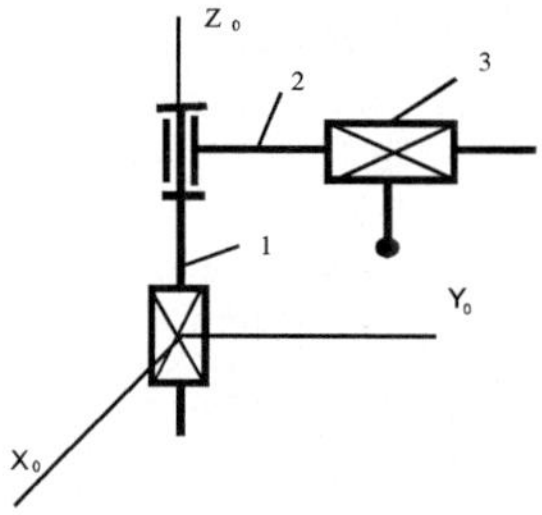	Angular velocity vectors and velocity vectors in translational movement—complanar vectors.One of the translational movementvelocity vectors coincides with the angular velocity vector

(R) and translational (T) kinematic pairs, then the most wide spread options are 6 combinations of three kinematic pairs: TTT, RTT, RRT, TRR, RRR, TRT (Table 1.2). In the kinematic chains formed of the listed above combinations, certain mutual orientation of the axes of rotational and guiding translational pairs must be ensured. Additional requirements to the three-DOF kinematic chains are absence of empty (inaccessible) areas in the servicing zone, and possibility of solving direct and reverse problem of kinematics. Out of functional properties obtained by CFG, it appears that any of obtained by CFG can control movement of the output point ***M*** in the space. Thus, the set problem of OMM synthesis can be solved by the transfer mechanism consisting of design schemes of two CFG. Mechanism of body rotation around the axis must not impose constraints on transferred by the CFG motions of the two body points. Hereinafter this mechanism is for our purpose is called the connecting link (*CL*). Reasoning from its intended use, the connecting link (Fig. 1.1) must have three-DOF articulation joint 1 (spherical pair) from one side and two-DOF articulation joint 4 (spherical pair with a pin) from the other. Structurally Connecting Link *(CL)* is produced with variable length and consists of 2 or 3 links forming prismatic kinematic pair, or it can have fixed length between the articulation joints centers, if the kinematic pair 2–3 is removed.

Centers of joints in the future we will call two points: the point M_1 of the intersection of the axes of three joints 1 and the point M_2 of the intersection of the axes of two joints 4. The output link (gripper) 5 has the ability to rotate by means of the drive 6 around the *CL* axis passing through the centers of the hinges. The required

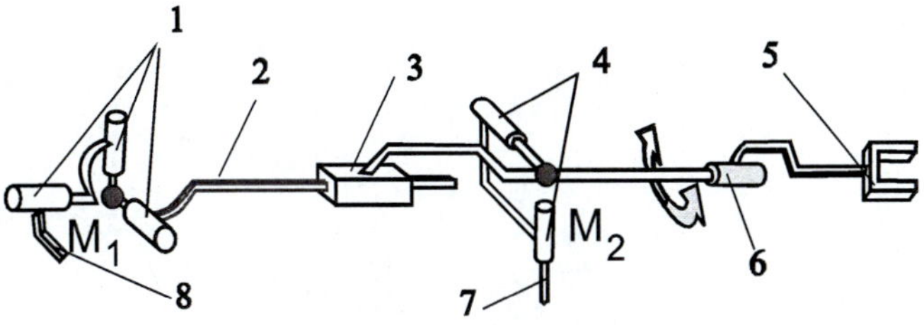

Fig. 1.1 Connecting link

orientation of the *CL* axis is achieved by changing the position of the centers of the hinges M_1, M_2 using the output links 7 and 8 of two CFG. It should be noted that in the *CL*, the output link 5 can structurally be placed also between the articulation joints centers. In order to move in space and orient the output link of manipulator of the robot in the space, it is enough to transfer joints centers in the *CL* using two CFG and turn the output link in relation to the *CL* axis. The specified motions can be obtained by various methods: by interconnecting two CFG and CL parallely as it is shown on the general diagram (Fig. 1.2a) or in series—parallely (Fig. 1.2b). As a result of the offered interconnection of the elements of the robot executive mechanism, manipulators are obtained consequently with parallel structures (Fig. 1.3) and hybrid manipulators with mixed (Fig. 1.4) structure (design scheme).

Seven-DOF manipulator with closed kinematic chain [17] shown on the Fig. 1.3 is a typical example of OPM. This manipulator is made of two CFG-1 according to the interconnection diagram (Fig. 1.2a).Connecting link 7 at the point *A* forms a spherical kinematic pair with the last link of CFG made up of links 1, 2, and 3.From the other side, the connecting link 7 forms a spherical kinematic pair with a pin together with the last link of the second CFG-1 made up of links 4, 5, and 6.*CL* imposes one constraint $AB = Const$. Thus, the working body 8 of the robot mechanism has 6 degrees of freedom including rotation motion with the help kinematic pair 4–8.In traditionally known manipulators of robots with the series connection of its links, usually the first CFG performs transfer motion, and the second CFG performs the orienting motion of the gripper.

For the synthesis of the design of various OPM robots, this paper offers to use the general diagram for forming manipulators with the closed kinematic chain that

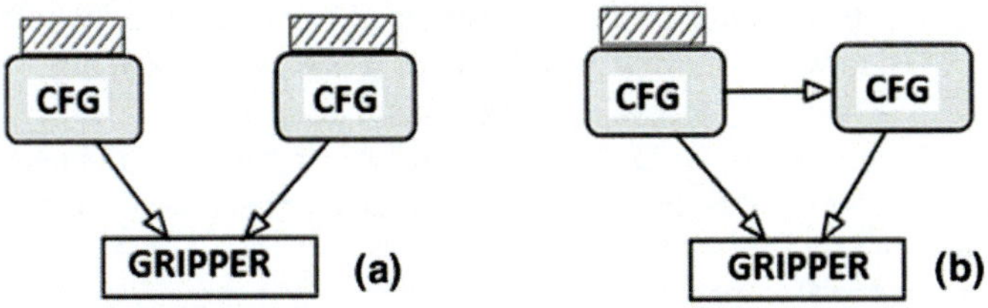

Fig. 1.2 General diagram for the synthesis of the one closed-loop parallel manipulators

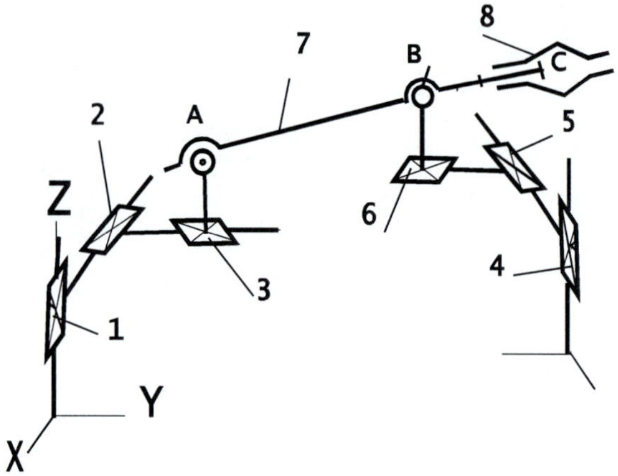

Fig. 1.3 Robot one-loop manipulator

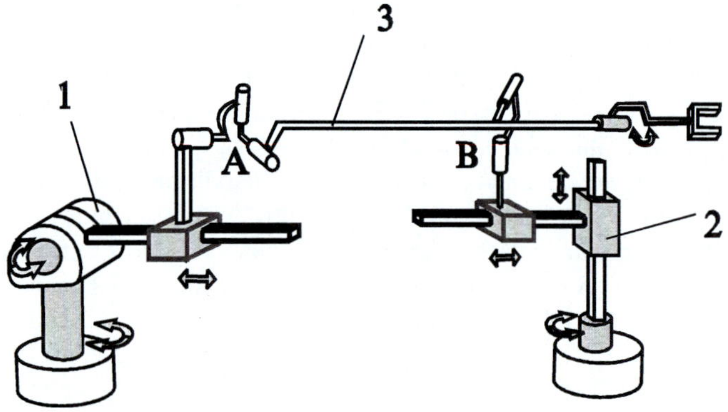

Fig. 1.4 OPMs of a Robot Made up of CFG: 1—CFG-3, 2—CFG-2, 3—CL

is obtained by the parallel connection of the CFG according to the general diagram given on the Fig. 1.2a.

The choice of this scheme of interconnection of the CFG and *CL* is based on that the OPM have significant advantages if compared to the manipulators with open kinematic chain. It is known that the serial connection of modules has lesser rigidity if compared to parallel connection of modules, and parallel structures have better characteristics of positioning accuracy, greater load-carrying capacity with the high parameters of rigidity and movement velocity of the working body.

The six CFG (Table 1.1) according to the general diagram (Fig. 1.2a) can make up 20 varieties of architecture for OPMs. If the Fig. 1.3 presents the architecture OPM made up of two CFG-1, then the Fig. 1.4 shows the architecture of a OPM made up of CFG-3 and CFG-2.

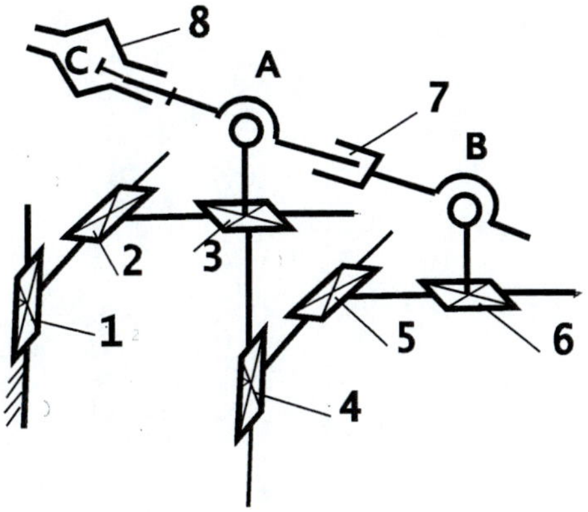

Fig. 1.5 One closed-loop hybrid manipulator with

When making parallel connection according to the scheme from Fig. 1.2a, as compound mechanisms, it is possible to use any of the above CFG or similar mechanisms with crossing or intersecting axes.

mixed structure

Hybrid manipulators with mixed structure made up according to the diagram on Fig. 1.2b can be of interest in some specific cases. This manipulators of robots are characterized by many drawbacks peculiar for mechanisms with open kinematic chain.

Design of the robot hybrid manipulator is formed according to the mixed connection scheme (Fig. 1.2b).In this case, the last link 3 of the first CFG made of links 1, 2, and 3 (Fig. 1.5) at the point *A* link 7 is connected with the help of spherical kinematic pair. The same link 3 with its other side is connected to the first link 4 of the second CFG made up of links 4,5, and 6.With its output link 6, the second CFG forms a spherical jointwith a pin at point *B* with the connecting link. The connecting link supporting the working body 8 has a variable length due to its prismatic connection. The connecting link length varies with the change in the position of links 3 and 6 of the CFG. The connecting link has 5 degrees of freedom. That is why, the working body 8 of the robot mechanism has 6 degrees of freedom, since the additional degree of freedom is provided by rotation in the kinematic pair 7–8.

It is possible to improve the dynamic characteristics of the manipulator without losing its functional capabilities when synthesizing the OPM from two incomplete CFG. If parallel planes are chosen, in which motion of output points is performed with the help of the incomplete CFG. These output points are to be connected to CL, and modified OPM is obtained. According to this principle, the two incomplete CFG-1 make up the mechanism of six-axis table [18]. Fig. 1.6a shows 3D model, and Fig. 1.6b shows the design scheme of the OPM obtained as a result of using the specified modification. This manipulator allows locating four actuators on the fixed base, which will improve its dynamic characteristics. As incomplete CFG, two-DOF CFG ensuring controlled motion of output point in the set plane can be used.

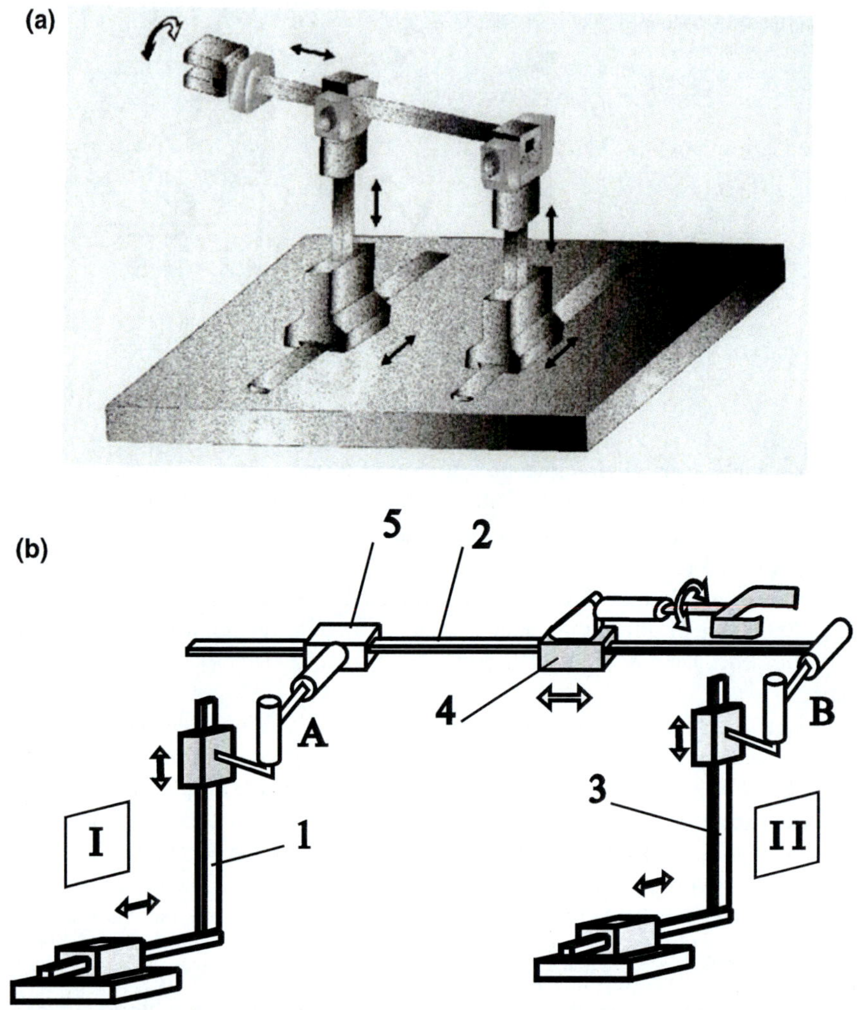

Fig. 1.6 Manipulator of the six-axis table

As it follows from the design scheme on Fig. 1.6b, the specified manipulator consists of two incomplete CFG-1 (1), (3) and *CL* (platform) 2.The two forward movements in each incomplete CFG are performed in parallel planes **I** and **II**, and forward movement of the link 4 relative to platform 2 is performed in the directions crossing planes **I** and **II**. Six-DOFOPM of this robot provides the manipulation object with six degrees of freedom. On the Fig. 1.6, the arrows mark the active actuators, in which the motion is implemented with the help of controlled drives. One of the peculiarities of this mechanism is also that its connecting link has the variable length due to its prismatic connection 2–5, and the gripping tool capable of rotational can be

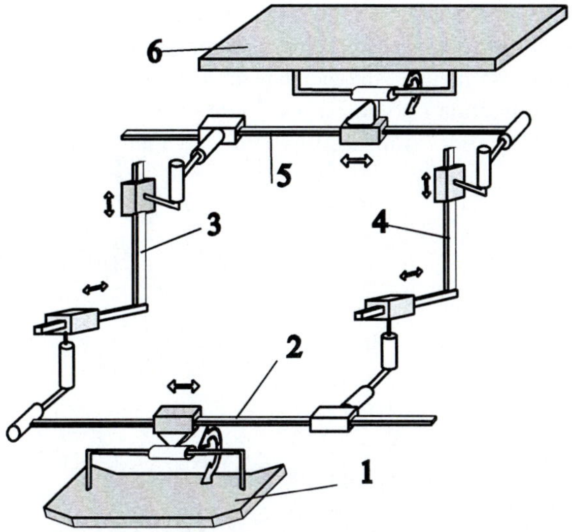

Fig. 1.7 OPM of the walking machine leg

placed both between the articulation joints centers *A* and *B*, and outside the centers, since they can be brought out in any direction along axis *AB*.

CFG connection according to the scheme of parallel connection can be used in the synthesis of the controlled mechanism of a walking machine leg [19]. Leg mechanism (Fig. 1.7) is made up of four functional elements: two parallelly connected kinematic chains 3, 4, two-DOF CFG-1, top 2 and bottom 5 connecting links. The leg mechanism obtained as a result of the synthesis allows implementing free (with six degrees of freedom) controlled motion of the casing 6 and foot 1 relative to each other. This becomes possible due to that in the transfer of, for example, the foot 1 with the help of brake devices (fixing mechanisms) excluding motion in articulation joints and actuators of the top connecting link 5.Then the actuators of each two-DOF mechanism move the articulation joints centers located on the bottom connecting link 2.

By additional forward and rotational motions relative to the longitudinal axis of the bottom CL, the carrying cushion 1 is moved with six degrees of freedom. In the position when the leg is supported by the foot, the brake devices exclude the motion in the articulation joints located on the bottom connecting link 2. Then centers of articulation joints of the top CL are moved by the set method using the actuators of the two-degrees-of- freedom mechanisms.

Additional forward movements with the help of the actuators relative to the top CL, allow moving the casing 6 orienting it simultaneously by the set method in the space. In Fig. 1.7 the active kinematic pairs are marked with the arrows. Thus, families of one-loop mechanisms of three types are obtained: with parallel connection of the CFG; hybrid and modified ones made of two incomplete CFG. Outstanding feature of the obtained system of one-loop multi-DOF mechanisms is that design of these robot mechanisms allows distributing the effect of the load between the two parallel kinematic chains. As a result, the bearing capacity of the mechanism increases, the rigidity of the structure and positioning accuracy of the output link are improved. At the same time, the mechanism weight decreases in general and the robot energy consumption also decreases. More over, as it will be shown below, for these robot architectures (in most cases), the direct and reverse problem of kinematic is solved unambiguously.

Chapter 2
Synthesis of Robot Multiloop Manipulators

Abstract Functionality of one-dimensional simplicial complexes, which are the graphs, do not allow to fully simulate the structure and relationships of spatial mobility mechanisms. Therefore, for modeling the design of mechanisms and kinematic pairs, it is proposed to use multidimensional, if compared to flow charts, simplicial complexes. Therefore, it is proposed to use the combinatorial topology methods for MPM architecture synthesis. Based on construction of the topological model of two interconnected bodies, the architecture of SHOLKOR multiloop parallel manipulator was obtained. Multi-link spherical, flexible and chain connections for MPM are offered.

Keywords Multiloop parallel manipulators · Topological model of interconnected bodies · SHOLKOR manipulator

2.1 Rationale for the Possibility of Using Methods of Combinatorial Topology to Modeling the Mechanisms and Manipulators Design

It is known that modern geometry and topology, due to the possibility of formalization and visibility, is an effective research tool in various fields of science. Moreover, today the computer geometry is widely coming into use, which allows visualizing complex mathematical and geometric objects. The geometry and topology deal with geometric shapes and objects implemented in the form of a set of points. An important concept used in the topology is the concept of homeomorphism defined as a mapping of one figure onto another, when forward and reverse mutually univocal and mutually continuous correspondence is set between the points of two sets. These properties are called topological properties. The concept of homeomorphism is convenient for identifying those properties of shapes, which do not change under the shape deformation. For example, homeomorphic shapes to the topological point of view are square and circle, cube, and tetrahedron, although when the metric properties are used, these are different objects. Here, a set of points formed by the mechanism during the possible movements of its links is represented as a topological space,

K. S. Sholanov, *Parallel Manipulators of Robots*, Mechanisms and Machine Science 92,
https://doi.org/10.1007/978-3-030-56073-7_2

the properties of which directly depend on the mechanism design. Therefore, for the analysis and synthesis of the mechanisms' and manipulators' design, modeling using topological spaces is applied [20–22]. At the same time, apparatus and methods of combinatorial topology are used [23–26].

2.1.1 The Basic Concepts of the Combinatorial Topology

Before the apparatus of the combinatorial topology is used, it is necessary to look at the basic concepts and definitions of the combinatorial topology. It is known that the combinatorial topology studies geometric shapes, breaking them in the right way into the simple shapes—simplexes. Those geometric shapes that can be in the appropriate way divided into simplexes are called polyhedra, while the scheme of breaking into simplexes is called a complex. Below are some basic definitions, which underlie the further conclusions.

Definition of the Simplex. Let $a_0, a_1, \dots, a_r$ be a system of independent points of the n-dimensional Euclidean space R^n, $(r \le n)$. A set A^r of points of the space R^n represented in the form of

$$x = \lambda^0 a_0 + \lambda^1 a_1 + \dots + \lambda^r a_r, \tag{2.1}$$

where $\lambda^0, \lambda^1, \dots, \lambda^r$ are real numbers satisfying the two following conditions:

$$\lambda^0 + \lambda^1 + \dots + \lambda^r = 1, \tag{2.2}$$

$\lambda^i > 0, i = 0, 1, \dots, r.$

This expression describes the *r-dimensional simplex* [24].

Hereinafter, the simplexes are referred to as A^r, B^s, C^n, while the expression of the form

$$A^r = (a_0, a_1, \dots, a_r)$$

means simplex $\boldsymbol{A}$ with dimensionality—r and vertices a.

If two simplexes A^r and B^s coincide, $A^r = B^s$, then their vertices also coincide with each other, of course, with precision to a degree, and in particular, $r = s$. Point x uniquely defines numbers $\lambda^0, \lambda^1, \dots, \lambda^r$ with the help of the correlation (2.1), unless the condition (2.2) is satisfied. This allows us to consider the number $\lambda^0, \lambda^1, \dots, \lambda^r$ satisfying the condition (2.1) as barycentric coordinates of the point $x \in A^r$. By definition, zero-dimensional simplex (a_0) consists of one point a_0. One-dimensional simplex (a_0, a_1) is a straight-line segment connecting points a_0 and a_1. Two-dimensional simplex (a_0, a_1, a_2) is a triangle with vertices a_0, a_1, a_2. Finally, three-dimensional simplex (a_0, a_1, a_2, a_3) appears to be tetrahedron with vertices a_0, a_1, a_2, a_3. Every two r-dimensional simplexes are homeomorphic, wherein their homeomorphism is performed by such reflection, in which the respective points have the same barycentric coordinates. Simplexes A and B of the Euclidean space R^n *are located correctly,* or if they do not intersect or their intersection $A \cap B$ is a face of each of the simplexes A and B. It is worth noting that each vertex of simplex A^r is its zero-dimensional face, and simplex A^r itself is its r-dimensional face. While the two faces of the simplex are always positioned correctly, it is known that the number of r-dimensional faces of the n-dimensional simplex is the number of combinations of C_{n+1}^{r+1} out of $n + 1$ elements along the $r + 1$. Hereinafter, the concept of oriented simplexes will be applied. One of the signs of the simplexes homeomorphism is their orientation. For this reason, on the face of a simplex, a direction is chosen, for example, anticlockwise. The said bypass has to be respected on all faces of the simplex under consideration. Then the simplex is considered to be oriented. It is necessary to explain concept of the complex that is basic for us. By the definition, finite set of C of simplexes of some of the Euclidean space R^n is called *geometric complex* or just a *complex,* if the following conditions are met:

1. Along each simplex A of the set C, the C also includes any face of simplex A.
2. Every two simplexes of the set C are positioned correctly.
3. Zero-dimensional simplexes of the complex C are called its vertices.

Maximal dimensionality of the simplexes from the C is called *dimensionality* of the C complex. Although the complex is the basic concept of the combinatorial topology, the real object of the research is not the complex but the topological space which is defined by it. In this paper, this is a set of geometric loci of bodies' points of a mechanical system with possible movements permitted by the constraints.

2.1.2 Rationale for the Use of Combinatorial Topology Methods

The structural and kinematic analysis of the kinematic pairs was conducted by various authors in most cases empirically. This was the cause of inconsistent results. Hitherto, for example, mathematically possible number of moving kinematic articulations of two bodies is not justified. Synthesis of mechanisms is based in most cases on engineering intuition and logic, as there is no sufficient scientific rationale for the choice of this or that kinematic pair in the mechanism design.

In solving a variety of applied problems in many branches of the science, recently topology methods are used increasingly. An example of applying the topology methods in solid mechanics is a large number of mechanical system studies conducted using all kinds of flowcharts. Currently, there is a certain experience available in applying the topological methods in the analysis of the mechanism design and complex systems [25, 26]. Topological apparatus was used for solving the mechanical system kinematic and dynamic problems in the paper [27]. However, the limited capabilities of one-dimensional simplicial complexes, which are the flow diagrams, do not allow to adequately simulate the structure of interrelations and mobility of bodies of complex mechanical systems such as multilink spatial mechanisms. Therefore, for modeling the design of mechanisms and kinematic pairs, it is proposed to use multidimensional, if compared to flowcharts, simplicial complexes. Thus, it is supposed to distribute some of the concepts and methods of the combinatorial topology to the analysis of the interconnected bodies' design.

In order to justify the possibility of applying the topology methods in the mechanical system research, the following *statements* are to be proved.

Statement 1. ***Space R formed from a set of points of the mechanical system with stationary holonomic constraints is a topological space.***

To prove this *statement*, let us consider the mechanism with open kinematic chain, which is shown in Fig. 2.1. This mechanism is composed of rack 0 and movable links 1 and 2. The links 0–1 form a one-DOF rotational pair, and the links 1–2 constitute the three-DOF spherical pair. Each of the links 1 and 2, respectively, will be associated with vectors ***AB*** and ***BC***. The system will be given the possible movement allowed by the imposed constraints. A set of positions of ends of vectors ***AB*** and ***BC*** in the space will be marked correspondingly via B and ***c***. A set of the points B and ***c*** form some area U so that $c \in U$. If G marks the area formed by the set of points ***c*** during the motion of point C allowed by the constraints, then the points of this area are $\boldsymbol{c} \cup G$. So that area S of space R is its basis, it is necessary and sufficient for any area G and any point c belonging to G, to find such area U of system S, which has $\boldsymbol{c} \cup G$. Space, for which basis S is introduced, is topological [24]. Thus, *statement* 1 is proved. As the example of the mechanism was chosen arbitrarily, the arguments of *Statement* 1 can be extended to any other set of links. Therefore, *Statement* 1 is true for any mechanical system with holonomic stationary constraints.

The elements of a topological space can also be generalized coordinates. In this case, the basis will represent an area formed by the variations of the generalized coordinates. The above restrictions of the constraints nature are conditioned by that only with holonomic stationary constraints, possible movements coincide with the set

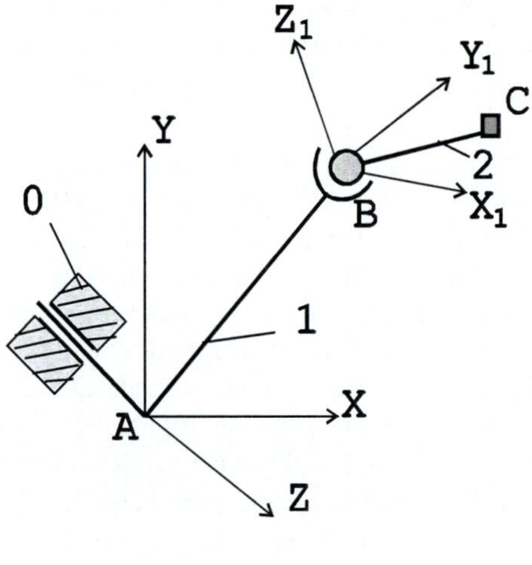

Fig. 2.1 Two-link spatial mechanism

of infinitesimal actual movement of the system that is compatible with the constraints. In this example, there is univocal correspondence between the set c and position of vectors ***AB*** and ***BC*** in case of possible movements of the system. Therefore, space T is the topological product of circle S_1 by two-dimensional sphere S_2, if

$$\left\{S_1(x_b, y_b)(x_b^2 + y_b^2 = l_1^2)\right\},$$

$$\left\{S_2(x_{1c}, y_c, z_c)(x_{1c}^2 + y_{1c}^2 + z_{1c}^2 = l_2^2)\right\},$$

$$T = [S_1 \times S_2],$$

where x_b and y_b are coordinates of point B in the coordinate system $AXYZ$ with the start at point A (Fig. 2.1); x_{1c}, y_{1c}, and z_{1c} are coordinates of point C in coordinate system $BX_1Y_1Z_1$ with the start at point C; l_1 and l_2 are lengths of links 1 and 2, respectively.

Statement 2. ***Change of the order of interaction of the mechanism links, i.e., the order of motion of the kinematic pairs of various mobilities, changes the topological space. The resulting topological space is not homeomorphic to the initial topological space.***

To prove *Statement* 2, the mechanism in Fig. 2.2 should be considered. This mechanism is made up of the same links as the mechanism in Fig. 2.1, with the only difference that the one-degrees-of-freedom and three-degrees-of-freedom kinematic pairs were interchanged. Repeating the arguments from the proof of *Statement* 1, topological space T_1 is obtained. Vector ***BC*** at any time is directed at a tangent to the two-dimensional sphere with the radius equal to ***AB***. In the differential geometry, it is known that there is no a continuous field of tangent vectors to a two-dimensional sphere.

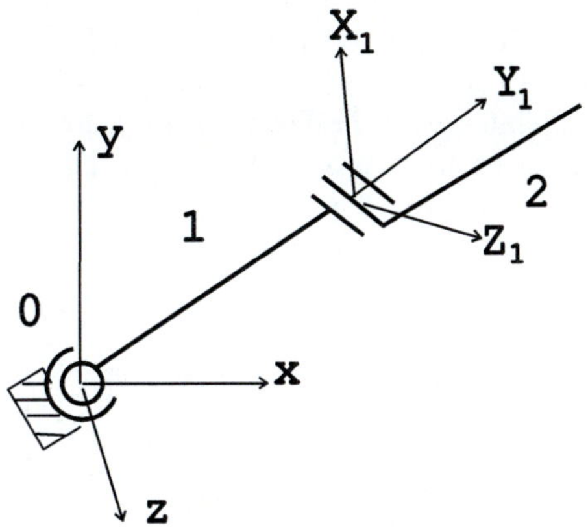

Fig. 2.2 Spatial two-link mechanism with reverse sequence of kinematic pairs

Therefore, space T_1 cannot be represented in the form of a topological product $S_2 \times S_1$ as space T. Hence, space T_1 is not homeomorphic to space T.

Thus, *Statement* 2 is proved, which has a significant importance in the theory of the mechanism design.

***Statement* 2 allows to judge the equivalence of kinematic chains through the research of their topological space homeomorphism [25]**. The given proof is the rationale that the mechanisms (Figs. 2.1, 2.2) are not equivalent to each other. It should be noted that the concept of equivalence underlies the classification of mechanisms.

2.2 Construction of a Topological Model of Two Interconnected Bodies

When modeling interconnected bodies of a mechanical system with the help of topological spaces, it is possible to solve a wide range of issues, if the apparatus of the combinatorial topology is used as a tool. The combinatorial topology allows studying topological spaces presenting them in the form of a simplicial complex, i.e., abstract scheme of tops and simplexes, while each simplex, in its turn, is a finite set of vertices. In order to build simplicial complexes that are models of interconnected bodies of the mechanical system, three topologies should be specified on the set of possible displacements of the system $X.\boldsymbol{\tau}_{1,}\boldsymbol{\tau}_{2,}\boldsymbol{\tau}_3$. Initial ones are the sets of X:A and B representing sets formed by points a and b of positions of two interconnected bodies of the system in case of possible movement of the system allowed by the imposed constraints, i.e., $a \in A$, $b \in B$, where $\cup A \subset X$ and $\cup B \subset X$. Topology $\boldsymbol{\tau}_1$ is given by setting binary relation $\mathrm{m} \in A \times B$ on the direct product of sets A and B, while, if $\mathrm{m} \in M$, then

$$X = \cup\{M : M \in \tau_1\}. \tag{2.3}$$

Pair $(X, \boldsymbol{\tau}_1)$ is a topological space, i.e., $T_1 = (X, \boldsymbol{\tau}_1)$. Here, $\boldsymbol{\tau}_1$ is a family of sets M that determine the correspondence of the set of all pairs (a, b) with possible displacement of two connected bodies relative to each other.

Similarly, on the direct product of the same sets A and B, topology $\boldsymbol{\tau}_2$ is determined using binary relation $c \in A \times B$. With $c \in C$, the topological space $T_2 = (X, \boldsymbol{\tau}_2)$ is built reasoning from that

$$X = \cup\{C : C \in \tau_2\}. \tag{2.4}$$

Here, $T_2 = (X, \boldsymbol{\tau}_2)$ is a topological space, where $\boldsymbol{\tau}_2$ is a family of sets C setting correspondence between the points of the set of all pairs (a, b), which impose constraints on each other. Topology $\boldsymbol{\tau}_3$ is determined by the binary relation $s \in A \times B$. The set of all points s belongs to the set S, i.e., $s \in S$. In this case, a topological space $T_3 = (X, \boldsymbol{\tau}_3)$, for which

$$X = \cup\{S : S \in \tau_3\}. \tag{2.5}$$

Topology $\boldsymbol{\tau}_3$ is a family of sets S made up of the same elements of the pairs (a, b) of the sub-sets A and B, for which $a = b$.

As a result, on one bearer space X, three topological products M, C, and S are determined. Association of topologies $\boldsymbol{\tau}_1$, $\boldsymbol{\tau}_2$, and $\boldsymbol{\tau}_3$ is the topology $\boldsymbol{\tau} = \boldsymbol{\tau}_1 \cup \boldsymbol{\tau}_2 \cup \boldsymbol{\tau}_3$. Thus, on the set of possible displacements of the mechanical system X, topology $\boldsymbol{\tau}$ is determined and the topological space $T_0 = (X, \boldsymbol{\tau})$ is established. On the other hand, sets A and B formed by points a and b of the positions of the two connected bodies with possible system movement allowed by the constraints are in their turn also topological spaces. Then the topological space T_0 can be determined as direct topological product of these spaces. The obtained conclusions prove that the set of positions of the mechanism interconnected bodies in case of possible displacement of the system is the topological space T that is sub-set of the topological space T_0.

2.2.1 Binary Relations Between the Vertices

The topological space that allows for triangulation (breaking into simplexes) is called a polyhedron. The topological properties of the polyhedron are set by its simplicial complex composed of simplexes. The simplest simplexes used hereinafter are zero-dimensional $A^0 = a_0$, one-dimensional $A^1 = (a_0, a_1)$, and two-dimensional simplex $A^2 = (a_0, a_1, a_2)$. Points belonging to the simplexes are called vertices. The topological spaces A and B during triangulation are presented in the form of simplexes A^r and B^s. Indexes of the simplexes r and s specify its dimensionality. Depending on the dimensionality r and s, 0,1,2-dimensional simplexes uniquely determine geometric elements: segment, plane, and space, respectively, up to a homeomorphism. Simplexes used below should actually be considered to be abstract and formed by these simplexes and simplicial complexes that are abstract complexes, as vertices of simplexes and complexes discussed hereinafter are marked ones.

When compiling the simplicial complexes of the interconnection of two bodies, the sets of positions, which are topological spaces A and B, are introduced according to the given topology of binary relations between two vertices of simplexes A^r and B^s.

Binary Relations and their Symbols:

1. If the topology $\boldsymbol{\tau}_1$ is given, then the vertices of the simplicial complex are connected by the face—*m*(*moving*) (Fig. 2.3a). Adjacency matrix in this case is denoted as $\boldsymbol{M}$.
2. In case, when topology $\boldsymbol{\tau}_2$ is given, simplicial complex vertices are connected by the edge—*c (connection)* (Fig. 2.3 *b*). Subsequently, simplexes' adjacency matrix is denoted as $\boldsymbol{C}$.
3. With the given topology $\boldsymbol{\tau}_3$, vertices of the complex are connected by the edge—*s* (*single*) (Fig. 2.3c). Adjacency matrix in this case is denoted as $\boldsymbol{S}$.

(a) (b) (c)

Fig. 2.3 Binary relations

From the kinematic point of view, the said binary relations-1,2,3 on the abstract schemes of the vertices of the simplex correspond to well-defined *states* of the kinematic parameters.

1. Parameters are independent and variable ($p_{\mathrm{ki}} = \mathrm{Var}$). In this case, the parameters are variable kinematic parameters.
2. Independent parameters p_{ki} are constants ($p_{\mathrm{ki}} = \mathrm{Const}$). In this case, the parameters determine the geometric values: linear dimensions, angles.
3. Parameters $p_{\mathrm{ki}} = 0$, i.e., parameters are dependent values.

These correspondences are crucial in constructing topological models of the kinematic pairs.

2.2.2 *Combinatorial Sum of Simplexes*

An important property of the simplexes is that the combinatorial sum C^n of the two simplexes A^r and B^s that are its opposite faces has dimensionality [24]

$$n = r + s + 1. \tag{2.6}$$

Adjacency of vertices A^r and B^s is determined strictly based on the condition that the resultant simplex C^n is correct and oriented.

Table 2.1 shows how the elementary 0-,1-,2-dimensional simplexes A^r, B^s (r, $s =$ 0,1,2) by consistent introduction of binary relations, for example, relation 2, generate the combinatorial sum-simplex C^n. Dimensionality of the simplex C^n is determined by equality (2.6).

If between the vertices of the opposite simplexes binary relations 1 are introduced, the comparative analysis of the combinatorial sum of two simplexes and kinematic properties of the two interconnected elements of a mechanical system establishes that the number of edges connecting the adjacent vertices of the simplexes A^r and B^s will be equal to the number of independent parameters that determine the relative position of the elements, the set of positions of which are simplexes.

Thus, for example, if $r = s = 1$ (Table 2.1), i.e., simplexes A^1 and B^1 present the set of straight-line positions, then the combinatorial sum of these simplexes has dimensionality $n = 3$. When forming the simplex C^3, adjacent vertices of the simplexes A^1 and B^1 will be connected by four edges. From the kinematic point of view, position of the two straight lines relative to each other in the space is also

Table 2.1 Building of simplexes

No	A^r	Image of A^r	B^8	Image of B^8	C^n	Image of C^n
1	A^0		B^0		C^1	
2	A^0		B^1		C^2	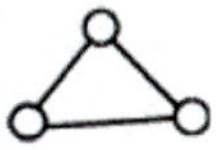
3	A^0		B^2		C^3	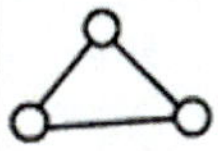
4	A^1		B^1	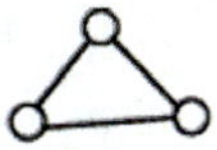	C^3	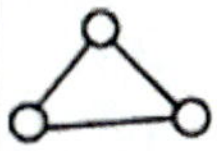
5	A^1		B^2	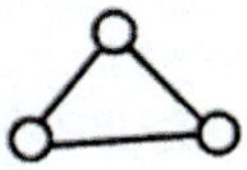	C^4	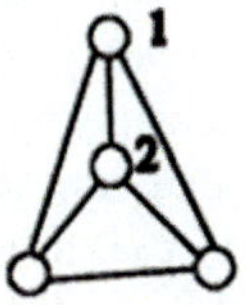
6	A^2	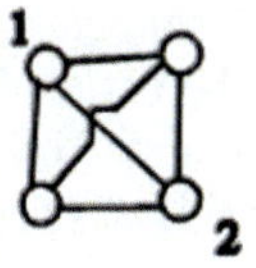	B^2	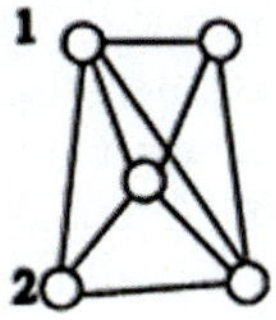	C^5	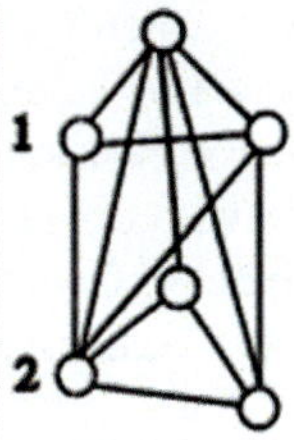

defined by four independent parameters (Denavit-Hartenberg [29] parameters for aligning axes X_{i-1} and X_i). Analysis of the data (Table 2.1) confirms the conclusion of Malyshev [28] that the point imposes three constraints on the body (see the simplex C^3 Table 2.1, line 3), and line imposes five constraints (see C^4). Table 2.1 shows the principle of forming simplexes with large dimensionality. However, the table allows for concluding the rule of decreasing dimensionality of the simplexes. Really, simplexes C^n form simplicial complexes, vertices of which are in general position. Although, if between vertices 1 and 2 of the complex formed by C^n, binary relation τ_3 is inserted and vertices are joint by the edge *s*, the simplex C^{n-1} (Table 2.1) is obtained, i.e., dimensionality of the simplexes is reduced.

In case there is a vertex, which includes or which emanates from the edge s as just one and the only, then this vertex can be removed and the dimensionality of the simplex decreases. Thus, a "compression" of the surface occurs, i.e., reduction of its dimensions. So, for example, if edges 1–2 in the simplex C^3 (line 4 in Table 2.1) is replaced by edge *s*, then vertex 2 will be the vertex from which only one edge *s* emanates. Therefore, this vertex is dependent and can be removed. In this case, the simplex C^3 turns into the simplex C^2 (line 2 Table 2.1). This action represents the surface compression process.

2.2.3 Construction of Simplicial Complex of a Kinematic Pair

For triangulation of the topological space T of a kinematic pair, topological spaces A and B are presented in the form of simplexes $A^2 = (a_0, a_1, a_2)$ and $B^2 = (b_0, b_1, b_2)$.

Given above topologies $\boldsymbol{\tau}_1$, $\boldsymbol{\tau}_2$, $\boldsymbol{\tau}_3$ (2.3–2.5) allow inserting the binary relations between the vertices of the opposite simplexes A^2 and B^2, or, which is the same, between the vertices of the simplicial complex formed by the combinatorial sum of the simplexes A^2 and B^2. In building, the simplicial complex of interconnected bodies, when a topology $\boldsymbol{\tau}_1$, is given, and vertices $a \in A^2$ and vertices $b \in B^2$ of the opposite simplexes are connected by the edge m. Such simplex simulates the interconnected bodies that have six degrees-of-freedom relative to each other. If in the simplex C^5 topology $\boldsymbol{\tau}_2$ is inserted, the vertices of the opposite simplexes will be joined by edges c. In this case, the simplex is a model of interconnected bodies in general position. In case, when the topology is inserted, the vertices of the opposite simplexes will be joined by edges s. In this case, the position space of the two bodies will have a smaller dimension, so that the dimension of the C^n simplex will be reduced by the number of edges s and become C^{n-s}. Simplicial complex (Fig. 2.4) composed of the combinatorial sum of simplexes A^2 and B^2 has six directed edges c (1–6) connecting adjacent vertices of opposite simplexes A^2 (I) and B^2(II). The edges are numbered

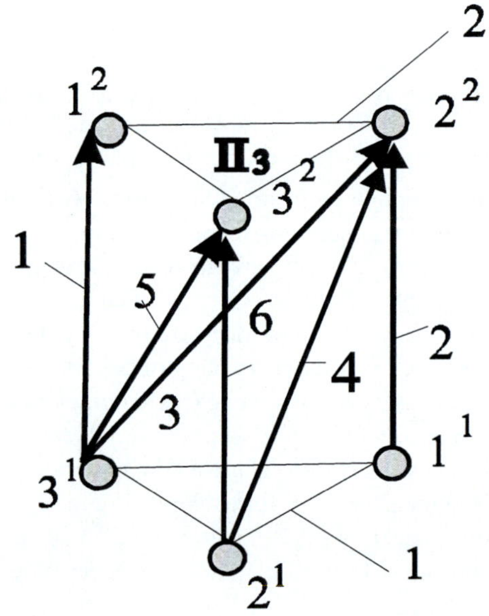

Fig. 2.4 Topological model of two interconnected bodies

according to certain rules associated with transformation parameters of i-coordinate system into $i-1$ system (see cl.), and edges are designated as p_k, where $k = 1-6$ is edge number. The vertices of the simplex are denoted by numbers equal to the number of edges emanating from it. The number of bodies (simplex) is denoted as the numeral in the exponent in the designation of the vertices. For example, 3^1 denotes the vertex of the simplex 1, from which three ribs emanate.

Vertices adjacency is established based strictly on condition that the resulting simplex was correct and oriented. While the simplex (1) corresponds to the $(i-1)$—body, and the simplex (2) to the i—body, i.e., on the simplicial complex, order relation is established from the simplex C^5. As is customary in combinatorial topology, the edge beginning is marked as *orig k*, and its end as *end k*. With the prescribed manner for the complex vertices shown in Fig. 2.4, there are *orig* $1 = 3^1$, *end* $1 = 1^2$, *orig* $2 = 1^1$, *end* $2 = 2^2$, etc. In Fig. 2.4, vertices of the simplexes 1 and 2 are joined by edges c. Therefore, the simplicial complex is a topological model of the interconnected bodies forming a fixed connection. It should be noted that the rule according to which the vertices of the simplex 2 are designated will be given hereinafter in the construction of the simplicial complex for the mechanical system. Hereinafter as well, as in Fig. 2.4), edges c in the simplicial complexes are replaced by arcs, i.e., by the directed edges. This is due to the fact that it is necessary to specify the coordinate system, which are superimposed during the transformation. To determine the position with the i-coordinate system relative to $i-1$ system, transformation parameters are determined by superimposing $i-1$ coordinate system with i-coordinate system. For this reason, in the simplicial complexes, the edges c are directed from the bottom $i-1$ simplex to the top i-simplex representing the i-body. Introduction of directed edges is associated with the fact that the order of the links' interconnection (as shown above) affects the results of the problem solution.

Depending on the topologies, the simplicial complex can analytically be described by square adjacency matrices $\boldsymbol{M}$, $\boldsymbol{C}$, $\boldsymbol{S}$ with dimensionality $[3 \times 3]$. Let us agree to consider the vertices, for example, in the matrix $\boldsymbol{C}$ to be adjacent, if the vertices of the bottom simplex are incident to the vertices of the top simplex and are connected by the edge c. In the case of vertices incidence, matrix element is equal to 1, otherwise to 0. Lines of the adjacency matrixes denote the vertices of the bottom simplex, and upper columns indicate the vertices of the top simplex.

For the simplex C^5 (Fig. 2.4), for example, adjacency matrix $\boldsymbol{C}$ looks as follows:

$$C_i = \begin{pmatrix} 0\,1\,0 \\ 0\,1\,1 \\ 1\,1\,1 \end{pmatrix},$$

where i is the number of interconnections. When the simplicial complex is described using the adjacency matrix $\boldsymbol{M}$, the matrix element is equal to 1, if the vertices of the simplex are incident, connected by edges m, and otherwise they are equal to 0.

For the analysis of possible elementary motions permitted by the interconnection of the two bodies, in the simplicial complex C^5, it is necessary to consequently

Table 2.2 Number of kinematic connections

m	1	2	3	4	5	**Σ**
C_n^m	6	15	20	15	6	**62**

replace edges c with edges m. From the kinematic point of view, this is equivalent to the serial release of the constraints and introduction of the mobility. The number of obtained as a result of this action complexes corresponding to the connections of the interconnected bodies with different mobilities is determined as the number of combinations of the total number of edges $n = 6$ by the number of edges m, i.e.,

$$X_\nu^\mu = \frac{\nu!}{\mu!(\nu-\mu)!}. \tag{2.7}$$

The number of possible options of mobility of two connected bodies found with the help of Eq. (2.7) is shown in Table 2.2. From Table 2.2, it follows that the number of, for example, one-DOF connections is six, and the number of all mobile connections, which can be formed by two solid bodies, is equal to 62. It should be noted that the data in the table does not indicate the number of kinematic pairs.

For example, the two known kinds of one-DOF kinematic pairs are the rotational and translational pairs. At that time, as the six one-DOF connections calculated using (2.7) and specified in Table 2.2 include forward motions along three independent vectors, i.e., three non-coplanar guides (guides may be tangential lines to the curves) and rotation around the three axes that are also arbitrarily oriented in the space.

When analyzing the mutual connection of two bodies and modeling them using topological spaces, the connection with each body of a certain axis is taken into account in the same way as when converting coordinate systems. Sometimes, it is needed to consider the point motion. For this reason, in the simplexes of the bodies forming a connection, it is necessary to highlight the elementary simplexes depicting point, straight line belonging to the respective bodies. Later, the condition is accepted that the vertices with index 1 in designating the vertices simulate a variety of positions of the point, and the vertices with indices 2 and 3 mean a set of direct positions belonging to the body under consideration. When modeling a particular interconnection of the two bodies in the form of a certain simplicial complex, it appears to be useful to break it into three-dimensional cores as shown in Fig. 2.5.

While the three-dimensional core K_1^3 (Fig. 2.5a) simulates position of the i-body relative to the point $i-1$ of the body, the three-dimensional core K_2^3 (Fig. 2.5b) simulates position of the straight line of the i-body relative to the straight line of the $i-1$-body.

The method of modeling the interconnection of bodies of a mechanical system using topological spaces and the reasoning carried out in this case is similar for any connections. Consider interconnection modeling method using simplicial complexes on the example of 3-DOF rotary or spherical joint (Fig. 2.6).

In the three-DOF rotational joint, there always is a point (3^1) connected with the $i-1$ body by the edges c (1,3,5), and the motion of the i-body relative to i–1 can be

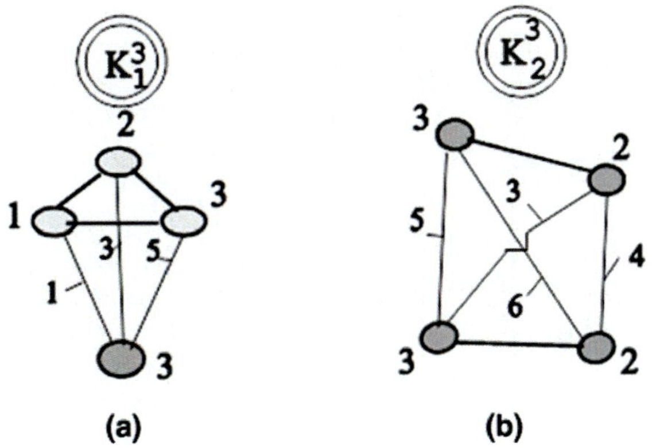

Fig. 2.5 Breaking of the simplicial complex

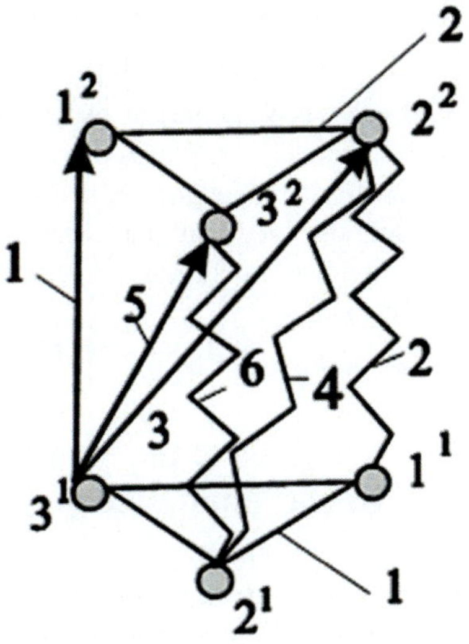

Fig. 2.6 Simplicial complex of the three-dimensional spherical joint of two bodies

decomposed into three rotational motions (depicted by the edges m (2,4,6)) around the three axes that do not lie in a single plane (Fig. 2.6). Any spatial motion of two bodies relative to each other can be uniquely represented as simplicial complex. This statement is confirmation below by the construction of three simplicial complexes. A simplicial model of the spatial displacement of some point of the second body relative to the first body can be constructed by freeing vertex 2^2 in the simplex (Fig. 2.2), herewith p_2,p_3, $p_4 = m$. The resulting simplex (Fig. 2.7a) is a simplicial model of the three-DOF movement in the space of some point of the body 2. If in the simplex (Fig. 2.2) we free points 2^2 and 3^2 from connections, we get a simplicial model of dimension 5, which shows movements with 5-DOF of a certain straight line 2^{2-32} of body 2 relative to body 1 (Fig. 2.7b). In the resulting simplex, by additional release from the connection of point 1^2 ($p_1 = m$), we obtain a model of the six-spatial motion of the i-th body relative to the i–1-th body. The simplicial model of free motion (with

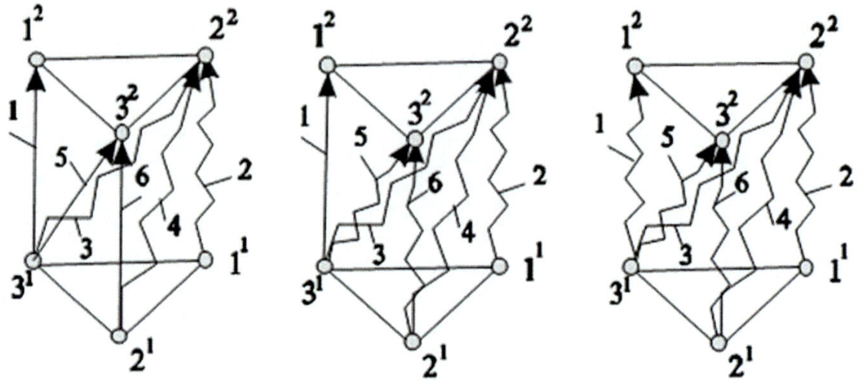

Fig. 2.7 Simplicial complexes of the 3-, 5-, 6-DOF joints of two bodies

six degrees-of-freedom) of two bodies relative to each other is shown in Fig. 2.7. Thus, simplicial complexes of various kinematic compounds can be constructed by choosing different edges *c* or *m* and their adjacency with the vertices of the complex. It should be noted that in all simplicial complexes, the number of edges m is equal to the number of degrees-of-freedom in the relative motion of two bodies forming a mobile joint.

2.3 Synthesis of the Structure of the Multiloop Parallel Robot Manipulator

The arguments given above (cl. 2.2.3) serve as a basis for constructing architecture of the multiloop parallel manipulator of robot in which the base is connected with the movable platform by several kinematic chains forming more than one loops.

Parallel manipulator of the *SHOLKOR* robot is built based on the topological model of interconnected bodies (Fig. 2.7c). For this, each edge *m* in the topological model of two interconnected bodies, is represented as a kinematic chain of the telescopic joint with ball pairs at their ends. The result architecture of the *SHOLKOR* robot (Fig. 2.8) [30] is obtained; it consists of the platform *II* connected by six kinematic chains B_1C_2, C_1A_2, C_1B_2, C_1C_2, A_1C_2, and B_1B_2 with another platform *I*. Platforms in motion relative to each other have six degrees-of-freedom, what is being modeled of the topological model of the interconnected bodies (Fig. 2.7c). Figure 2.8 shows the connecting kinematic chains (CKC) 1–6 with active actuators varying the chain length. CKC 1-6 with the help of multilink spherical joints [31] are connected to the bottom I and top II platform. While the CKC form with the platforms: at the nodes A_1, A_2—two-link spherical joints; at the nodes B_1, B_2—spherical three-link joints of the links I,5,3 and II,4,6, respectively (Fig. 2.9a); at the nodes

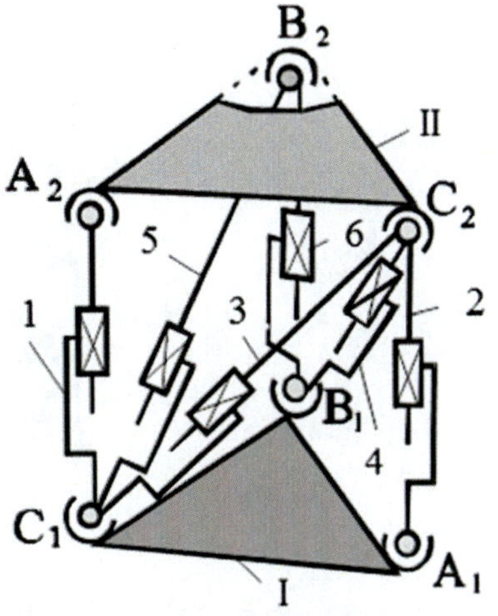

Fig. 2.8 Architecture of the SHOLKOR manipulator

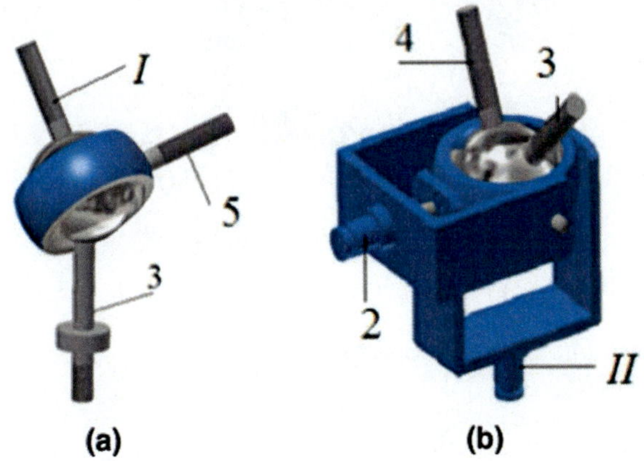

Fig. 2.9 Multilink spherical joints

C_1, C_2—four-link spherical joints of the links I,1,3,5 and II,2,3,4 (Fig. 2.9b). In the three-link spherical joint (Fig. 2.9a), the link 1 has a hollow spherical head that with its outer surface forms a spherical joint with the spherical cage of the link 5, and with its inner surface forms the spherical joint with the spherical head of the link 3.

The four-link spherical joint (Fig. 2.9b) is made up of a Cardan joint with forks II and two, in which the crosspiece is made in the form of a spherical cage that forms spherical connections with links three and four similar to the three-link spherical connection shown in (Fig. 2.9a). In six-DOF parallel manipulator according to the scheme of the links, interconnection of the parallel *SHOLKOR* manipulator, given above spherical multilink joints of two, three, or four links, can be replaced by chain connections [32] made of the two chain links 7–8 and 9–10; three chain links 7–8, 9–10, and 11–12; and four chain links 7–8, 9–10, 11–12, and 13–14, which are shown in Fig. 2.10a–c, respectively. It should be noted that Fig. 2.10a, b shows that the link 8 of the connection forms the rotational pair with the platform (bottom or top). In

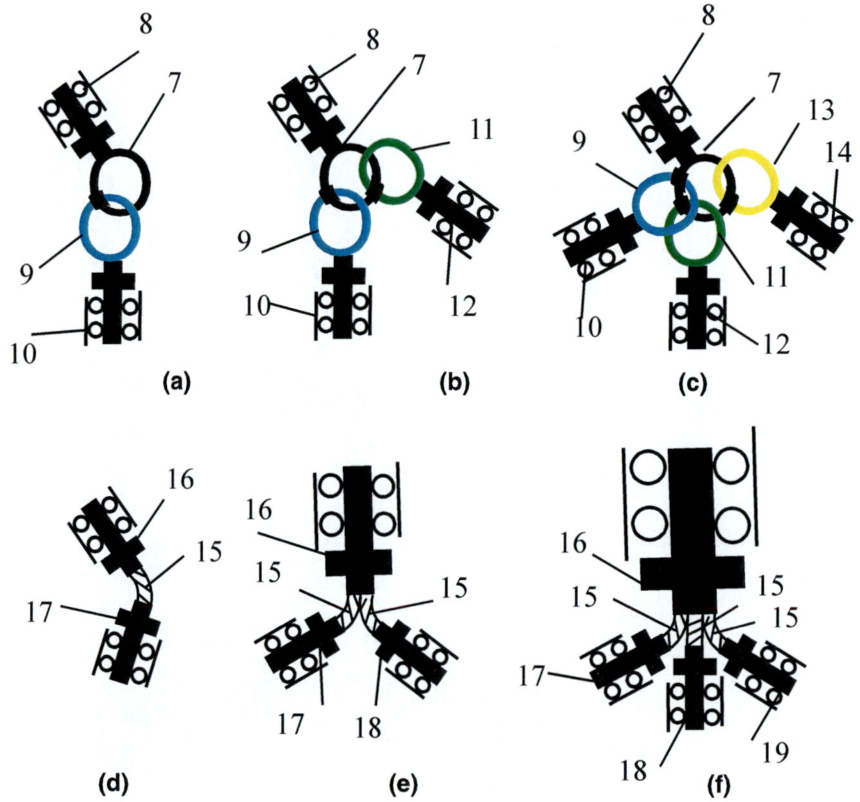

Fig. 2.10 Multilink connections

the form of spherical cage that forms a spherical joint with the links, three and four are similar to the three-link spherical joint shown in Fig. 2.9 a. Other links 10, 12, and 14 form rotational pairs with CKC.

Spherical multilink joints of two, three, and four links may also be replaced, respectively, by flexible connections [32] with the help of cable (rope) 15 connecting the two links 16 and 17 (Fig. 2.10d); three links 16, 17, and 18 (Fig. 2.10e) connected by two cables 15; and four chain links 16, 17, 18, and 19 (Fig. 2.10f) connected by three cables 15. In this case, the link 16 forms rotational pair with the top or bottom platform, and the links 17, 18, and 19 form rotational pairs with CKC. As shown in Figs. 2.9 and 2.10, various options of multilink spherical, chain, and flexible connections allow for structural implementation of the parallel manipulator of the *SHOLKOR* robot. Figure 2.11a, b shows demonstration models of the manipulators of the *STEWART-GOUHG* and *SHOLKO R* robots that have elastic multilink connections.

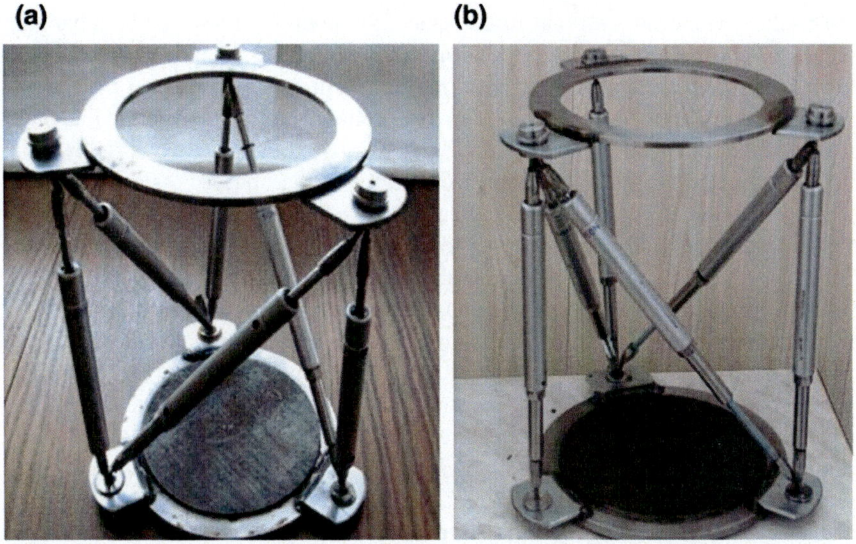

Fig. 2.11 Demonstration models of multiloop manipulators of robots: *STEWART-GOUHG* **a** *and SHOLKOR* **b** *platform*

2.3.1 Comparative Analysis of the STEWART-GOUHG and SHOLKOR Platforms

In existing multiloop robotic manipulators similar to the *STEWART-GOUHG* platform, there is a significant effect of the mobility of the links of adjacent kinematic chain on each other. Therefore, to perform a predetermined moving of the output link in these robots, it is necessary to provide coordinated motions of the different kinematic chains in the actuator kinematic pairs. These circumstances substantially limit the field of application of the existing multiloop parallel robots.

In order to make a comparison of two multiloop robots with *STEWART-GOUHG* platform architecture and *SHOLKOR* robot, the demo models of the manipulators of these platforms were produced (see Fig. 2.11). As a result, the following is established:

First, each of the six actuators of the *SHOLKOR* robot can move independently of each other. For example, it is possible to change the length of one *CKC* using the actuator without changing the lengths of the other *CKC*. The top platform will take a strictly defined position relative to the bottom platform. In the *STEWART-GOUHG* platform, independent changes in the connecting chains length are impossible.

Second, by changing the length of a certain set of *CKC* in a predetermined way in the model of the *SHOLKOR* robot, the required spatial position of the mobile platform or its simplest motions relative to the bottom platform can be obtained. Thus, for example, the movement of the platform *II* relative to the platform *I* can be performed by positioning some straight line B_2C_2 in the space (Fig. 2.8). Thus, first to the predetermined point in the space, the node C_2 is moved via changing lengths of connecting chains 2, 3, and 4; then the node B_2 is moved by changing the lengths of the connecting chains 5 and 6. Positioning of the top platform is completed relative to the bottom platform by rotational around the straight line B_2C_2 of the node A_2 by changing the length of the chain 1. From the above, it follows that the spatial positioning of the platform *II* relative to the platform *I* can be divided into three consecutive phases: movement of the node C_2, and then nodes B_2 and A_2. These transfers are obtained as a result of analyzing of simplicial complexes (Fig. 2.7) described above. Thus, the above-presented simplicial complexes with different combinations of the edges *c* and edges *m* are the models of spherical motion of the platform in the space (Fig. 2.6), moving some point of the platform in the space (Fig. 2.7a) and moving some straight line of the platform in the space (Fig. 2.7b). The demo model confirmed that these motions can be implemented by the *SHOLKOR* robot. Thus, *SHOLKOR* robot architecture enables targeted control of the top platform motion that follows from the topological model (Table 2.1), and this is confirmed by the demonstration model. *STEWART-GOUHG* platform, unfortunately, cannot enable such known motions as the spatial positioning of some point of the movable platform and the spherical motion of the platform or platform rotation around some axis.

Part II
Analysis of the Design of Robot Parallel Manipulators

Chapter 3
Determination of the Number of Degrees-of-Freedom of Mechanisms Using Reference System Imaging

Abstract It is known that until now there is no any unified theory and methods obtained for determining the number of degrees of freedom, which would be applicable to any mechanism. OPM and MPM mechanisms are spatial mechanisms. Offered herein theory of spatial mechanism design analysis is also relevant for robot parallel manipulators. The design analysis theory is based on the zero parameters method. This method uses transformation of coordinate systems related to the links with transformation parameters equal to zero (zero parameters), which characterize the features of links configuration. Examples of zero parameters application are provided. In particular, through the example of the Bennett mechanism, the use of zero parameters method is demonstrated for the synthesis of mechanisms without dependent links.

Keywords Number of degrees of freedom · Transformation parameters · Zero parameters · Dependent links

3.1 State of the Theory of Closed Mechanisms Design Analysis

One of the tasks in the mechanisms design analysis is determination of the number of degrees-of-freedom (NDF). Since the mechanisms of one-loop and multiloop robots are multi-degree-of-freedom spatial mechanisms with closed kinematic loops, the task of determining the NDF for these robots and in general for the spatial closed mechanisms is topical. When solving the kinematic problems, the number of degrees-of-freedom of a mechanism determines the number of generalized coordinates (rim variables), and it also determines the dimensionality of the configurational space (for robots). In dynamics and control of robot mechanisms, the number of independent dynamical equations is equal to NDF (for a holonomic system with absolutely solid bodies). Therefore, it is important to have an opportunity to determine the NDF for any one-loop, multiloop mechanisms including robot manipulators. However, analysis performed by Gogu [33] for the studies published during the last 150 years showed that until present a theory and method of determining NDF applicable to

K. S. Sholanov, *Parallel Manipulators of Robots*, Mechanisms and Machine Science 92,
https://doi.org/10.1007/978-3-030-56073-7_3

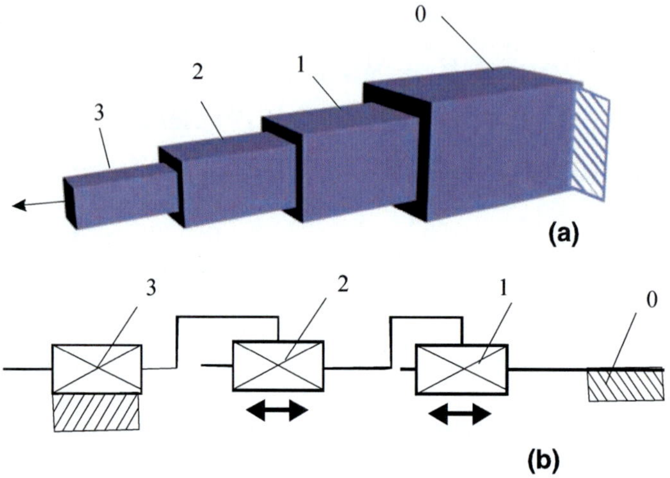

Fig. 3.1 Telescopic mechanism

any mechanism had not been obtained. Earlier, formula for NDF that included the number of dependent connections was obtained in the study [34]. However, method for determining the dependent connections given in this paper is applicable only to mechanisms with rotational pairs and supposes for the use of standard multitude of positions of rotational joint axis. Analysis of the researches studying the determination of the number of degrees-of-freedom shows that they can be divided into studies based on algebraic methods [35, 36] and methods that use theory of screws [37].

Analysis of the mobility equation shows that they include such values as number of moving links, number of joints with different degrees-of-freedom, number of loops, and number of dependent (or independent) constraints. At the same time, discrepancy between the actual mobility and mobility determined by this or that formula mainly depends on the method of determining the number of dependent constraints, while these methods must be applicable to any mechanisms and be convenient in use.

It is obvious that the closing of a kinematic chain often changes the number of conditioned by the joints constraints that existed before the closing. This is due to the peculiarities of the relative position of joints.

For example, open kinematic diagram of the telescopic antenna mechanism (Fig. 3.1a) has three moving links ($n = 3$) and three one-degree-of-freedom translational joints ($C_5 = 3$). In this case, there are 15 constraints conditioned by the joints. According to the Somov-Malyshev [28] formulas, the number of degrees-of-freedom is equal to

$$W = 6n - \sum_{i}^{5} iC_i = 6 \times 3 - 5 \times 3 = 3.$$

If the last link 3 is fastened, then a closed mechanism shown in Fig. 3.1b is obtained. For this mechanism with closed kinematic chain, using the same formula, the following is obtained:

$$W = 6 \times 2 - 5 \times 3 = -3.$$

As it is seen, result obtained from the formula does not correspond to the apparent thing that links 1 and 2 (Fig. 3.1b) can move independently from each other, i.e., the mechanism has two degrees-of-freedom and $W = 2$. The discrepancy between the actual and calculated NDF is associated with the fact that when closed, the number of constraints in a loop in a closed mechanism changes and became 10. This example showed that when the kinematic chain is closed, the number of constraints decreased for the value of $S = 5$. Value (S) hereinafter is referred to as the number of dependent constraints.

Thus, the dependent constraints appear as a result of closing the kinematic chain, while the number of dependent constraints depends on the relative position of links and joints, i.e., the orientation and positioning of the reference system axes. In this regard, this study seeks to obtain convenient for practical use and versatile method for determining the number of dependent relations for further calculating NDF of the closed mechanisms, including robot manipulators. At the same time, the task is to use the methods of imaging the reference systems that are commonly used in kinematics, as well as dynamic equations.

3.2 Reference System and Transformation Matrix Parameters

Hereafter, the analysis of the structure of closed complex mechanisms (manipulators) uses the generally applied in the kinematic systems of coordinates associated with the links. However, in the modern literature on the kinematics of mechanisms, different approaches to the interpretation of a number of concepts are often found. In this regard, in order to avoid ambiguity in the interpretation and terminology, let us briefly discuss the concepts of kinematics used herein. Under the reference system, a set of four vectors allowing to determine the position and orientation of the coordinate system is understood; this coordinate system is associated with a localized body relative to the absolute (inertial) coordinate system [38]. The reference system can, in general, be used to describe one coordinate system relative to another. Description of a reference system consists of a description of the position and orientation of the localized system. At the same time, there are various ways of the three-element description as the position and orientation. For example, the body position may be defined as a radius vector of the coordinate system related to the localized body relative to the absolute coordinate system and given in the form of three elements: coordinates of the beginning of the localized system of coordinates on the three axes

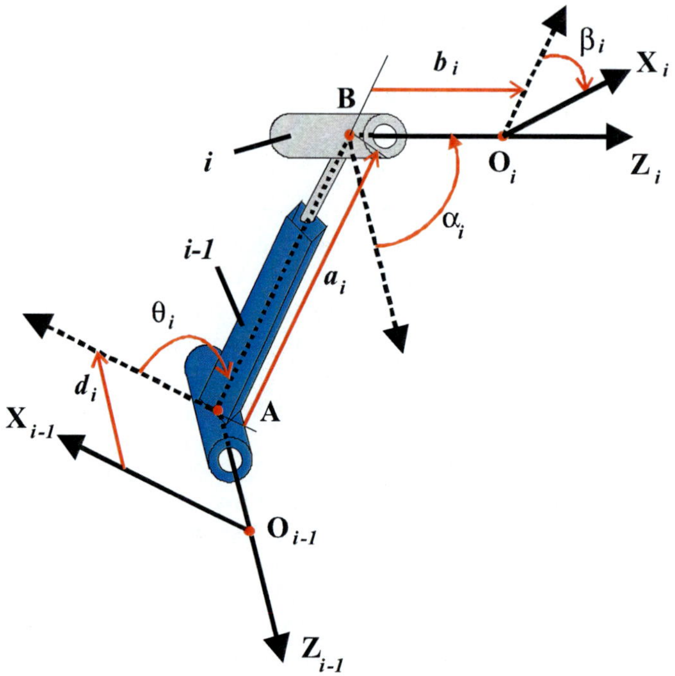

Fig. 3.2 Transformation parameters

of the absolute coordinate system. Thus, the description of the reference system uses six mutually independent elements. From kinematics, it is known that the position of a mechanism and relative position of the links can be determined by using the homogeneous transformation matrix A_i^{i-1}. Homogeneous matrix A_i^{i-1} of transformation of $i-1$ system of coordinates in the i-system allows determining the position of i-link relative to i–1 system of coordinates. Position of the mechanism links (when using A_i^{i-1}) is determined with the help of the traversal of the mechanism loop with the coordinate trihedron. The main actions in traversing are transformation of the system of coordinate described by the matrix A_i^{i-1}. Matrix elements A_i^{i-1} are transformation parameters or trigonometric functions from the transformation parameters. *Transformation parameters are values equal to the linear and angular displacements used when combining the coordinate systems associated with the links forming the movable connections (joints).*

Figure 3.2 shows the transformation parameters d_i, θ_i, a_i, α_i, b_i, and β_i, used in combining the coordinate system $O_{i-1}X_{i-1}Y_{i-1}Z_{i-1}$ with the coordinate system $O_iX_iY_iZ_i$.

In this paper, the transformation parameters are used as values determining the position of the links relative to each other.

Further, a more detailed description will be given to the role of the transformation parameters. It is worth noting that now the specific transformation parameters are not determined for homogeneous transformation matrices A_i^{i-1}, and matrix equations are not solved as is customary in the kinematics. Below it will be shown that the number of dependent constraints can be determined indirectly by analyzing the transformation parameters.

3.3 Binding of the Reference Systems to the Links

A peculiarity of the method of zero parameters in analyzing the mechanism design is that according to a particular rule that each of the links $i-1$, i forming $(i-1-i)$ joint ($i = 1, \ldots, n$, n if the number of moving links), binds to $O_{i-1}X_{i-1}Y_{i-1}Z_{i-1}$ and $O_iX_iY_iZ_s$ stems of coordinates, respectively. Fixed link (support) is indicated by the zero digit (0).

The main condition when choosing the coordinate systems is that all the kinematic (variables) values characterizing the position of the links, as well as geometric constant values (link dimensions, twist angles, etc.) are included in the number of transformation parameters.

When choosing the coordinates system, it is recommended to

- Direct axes Z_{i-1} along the axes of cylindrical, screw, prismatic, rotational ($i - 1 - 1$) joints, or perpendicular to the planes, in which there are slots for the pins of the moving spherical $(i-1-1)$ joints.
- Choose the beginning of the system of coordinates $O_{i-1}X_{i-1}Y_{i-1}Z_{i-1}$ in the centers of the spherical $(i-1-1)$ joints. Choose the direction of the axis X_0 at random. Direction of the axis X_i is chosen perpendicular to the plane, where axes Z_{i-1} and Z_i are located if these axes are crossed, or perpendicular to the axes Z_{i-1} and Z_i. In the spherical kinematic pair with a pin, axis X_i is directed along the axis pin.
- Direct axis Y_i so that the right system of coordinates is obtained (hereinafter, axes Y_i are not shown, as their position, when axes X_i, Z_i *are set,* is determined unambiguously).

Signs of the correctly chosen coordinate systems are that

- Combining of the coordinate systems is performed using a minimum number of the transformation parameters;
- when the position of the links varies, only variable transformation parameters called articulation parameters are changed;
- when the change of the link position does not change the transformation parameters characterizing the geometrical values (dimensions, twist angles) called link parameters.

3.4 Transformation of the Coordinate Systems Related to the Links

Given the important role assigned in this method to the transformation parameters, further let us focus on how to determine them. Transformation parameters should be determined by combining $i-1$ system with i-system of coordinates, according to the results of successive angular and linear displacement through the straight line AB that is common perpendicular to the axis Z_{i-1} and Z_i (it is known that any two lines arbitrarily positioned in the space have a single common perpendicular between them).

In order to determine the transformation parameters for the ($i-1-$i) joint, successive steps are to be performed for combining the system of coordinates $O_{i-1}X_{i-1}Y_{i-1}Z_{i-1}$ with the system of coordinates $O_iX_iY_iZ_i$ (Fig. 3.3).

Step 1 A shift of the i–1-coordinate system along the axis Z_{i-1} for the distance d_i from the center of coordinates O_{i-1} to the point A of the crossing the axis Z_{i-1} with the straight line AB, while axis X_{i-1} takes a new position $X_{i-1}^{(1)}$ (Here and below, the superscript in parentheses indicates the number of new positions of the coordinate system).

Step 2 A turn for the angle of θ_i around the axis Z_{i-1} until the axis $X_{i-1}^{(1)}$ is directed along the line AB (new position of the axis X_{i-1} is denoted as $X_{i-1}^{(2)}$).

Step 3 Shift for the value of a_i of the shortest distance between the axis Z_{i-1} and Z_i (in this case, the axis Z_{i-1} takes the position of $Z_{i-1}^{(3)}$).

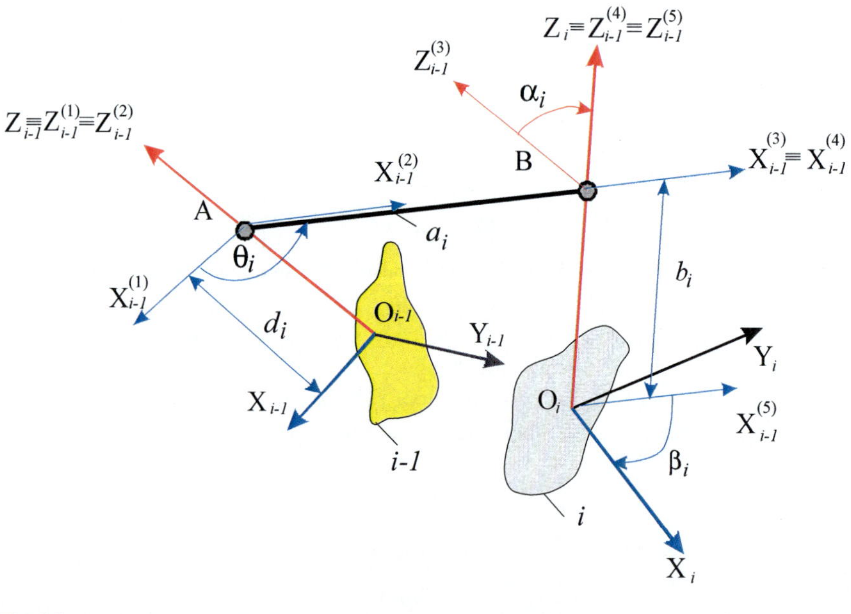

Fig. 3.3 Positions of the coordinate system and six transformation parameters

Step 4 A turn for the angle of α_i around the axis $X_{i-1}^{(3)}$ that is perpendicular to the axes Z_{i-1} and Z_i, till the directions of the axes $Z_{i-1}^{(3)}$ and Z_i *coincide*.

Step 5 Transfer along the axis Z_i for the distance of b_i (through $X_{i-1}^{(5)}$ it marks the new intermediate position of the axis X_{i-1}).

Step 6 Turn around the axis Z_i for the angle of β_i, until a complete match of the coordinate system $O_{i-1}X_{i-1}Y_{i-1}Z_{i-1}$ with coordinate system $O_iX_iY_iZ_i$.

Thus, the relative position of the links forming (i–1–i) joints is uniquely determined by six transformation parameters: three linear movements d_i, a_i, b_i along the non-coplanar axis Z_{i-1}, AB, Z_i and three angles of rotation θ_i, α_i, β_i around the same axes. Cross headings of the axis Z_{i-1}, AB, Z_i will be denoted, respectively, by unit vectors $\vec{k}_{i-1}, \vec{e}_i, \vec{k}_i$. To analyze the kinematical pair (KP) of the mechanism formed by i and $i-1$ links, hereinafter the transformation parameters will be denoted by $\vec{p}_{ki}$, where index $k = 1 \div 6$ is the number of the transformation parameter; index $i = 1 \div N$, where N is the number of KP in the mechanism.

With the accepted designation

$$\vec{p}_{1i} = d_i\vec{k}_{i-1}, \vec{p}_{2i} = \theta_i\vec{k}_{i-1}, \vec{p}_{3i} = a_i\vec{e}_i, \vec{p}_{4i} = \alpha_i\vec{e}_i, \vec{p}_{5i} = b_i\vec{k}_i, \vec{p}_{6i} = \beta_i\vec{k}_i. \quad (3.1)$$

It should be noted that in those cases when this is possible, the method of choosing and transforming the coordinate systems according to Denavit-Hartenberg can be used. As is known, in this case, the four transformation parameters are used.

3.5 Analysis of the Transformation Parameters

Transformation parameters d_i, θ_i, a_i, α_i, b_i, and β_i may be variable or constant values, or may be equal to zero. Thus, as previously indicated, the variables (kinematic values) are joint parameters, while the constants (geometric values) are link parameters.

Parameters equal to zero are particularly noteworthy. These parameters characterize the peculiarities in the relative position of the coordinate systems $O_{i-1}X_{i-1}Y_{i-1}Z_{i-1}$ and $O_iX_iY_iZ_i$. Thus, null parameters characterize peculiarities in joints configuration. Table 3.1 and Fig. 3.4 show (based on the designation of coordinate systems in Fig. 3.2) peculiarities in the arrangement of the coordinate systems caused by the equality to zero of this or that transformation parameter.

3.6 Rationale for the Method of Determining the Number of Dependent Constraints

Let us consider mechanical system with holonomic stationary constraints of N moving links ($i = 1, \ldots, N$) [39]. All forces affecting each i-link, including forces of

Table 3.1 Cases of occurrence of zero transformation parameters

Parameters amounting zero	Peculiarities in the arrangement of the coordinate systems $O_{i-1}X_{i-1}Y_{i-1}Z_{i-1}$ and $O_iX_iY_iZ_i$
d_i	The center of the coordinate system $O_{i-1}X_{i-1}Y_{i-1}Z_{i-1}$ coincides with the point A at the crossing of the axis $O_{i-1}Z_{i-1}$ a common perpendicular AB (Fig. 3.4a)
θ_i	Axis $O_{i-1}X_{i-1}$ is directed parallel to the line AB or perpendicular to the axis Z_i (Fig. 3.4b)
a_i	Axes $O_{i-1}Z_{i-1}$ and O_iZ_i intersect (Fig. 3.4c)
α_i	Axes $O_{i-1}Z_{i-1}$ and O_iZ_i are parallel (Fig. 3.4d)
b_i	The center of the coordinate system $O_iX_iY_iZ_i$ coincides with the point B of the crossing of the axis O_iZ_i with common perpendicular AB (Fig. 3.4e)
β_i	Axis O_iX_i is directed parallel to the straight line AB (Fig. 3.4f)

inertia, will be presented in the form of three component forces F_i^l and three pairs of forces M_i^l $(l = 1, 2, 3)$, directed along the axes $\vec{k}_{i-1}, \vec{e}_i, \vec{k}_i$. Then all the forces and moments can be represented as vectors

$$\vec{\bar{Q}}_{1i} = F_i^1\vec{k}_{i-1};\ \vec{\bar{Q}}_{2i} = M_i^1\vec{k}_{i-1};\ \vec{\bar{Q}}_{3i} = F_i^2\vec{e}_i;\ \vec{\bar{Q}}_{4i} = M_i^2\vec{e}_i;\ \vec{\bar{Q}}_{5i} = F_i^3\vec{k}_i;$$
$$\vec{\bar{Q}}_{6i} = M_i^3\vec{k}_i \tag{3.2}$$

The general dynamic equation for the given system on a virtual displacement can be recorded as a system of equations

$$\sum_{k=1}^{6} \bar{\bar{Q}}_{ki}\delta\bar{\bar{p}}_{ki} = 0 \quad (i = 1, \ldots, N), \tag{3.3}$$

where $\bar{\bar{Q}}_{ki}$—row vector whose elements are the vectors $\bar{Q}_{ki}$ (3.2) and $\delta\bar{\bar{p}}_{ki}$—column vector with elements in the form of vectors (3.1). Matrix Eq. (3.3) is equivalent to the system i of the linear equations (i = 1, …, N) obtained as a result of matrix multiplication, namely,

$$\delta d_i F_i^1 + \delta\theta_i M_i^1 + \delta a_i F_i^2 + \delta\alpha_i M_i^2 + \delta b_i F_i^3 + \delta\beta_i M_i^3 = 0. \tag{3.4}$$

This system of equations describes the linear space of dimensionality 6, as the rank of the coefficient matrix along the columns does not exceed six. If all the links of the kinematic chain are connected to each other rigidly by putting six constraints on each joint, all the equalities (3.3) are satisfied, and the equations of the system will have a rank equal to six. Consequently, the system of equations describes the space of the constraints, dimensionality of which is less than or equal to six. Change of the dimensionality of the space constraints, i.e., the number of independent constraints,

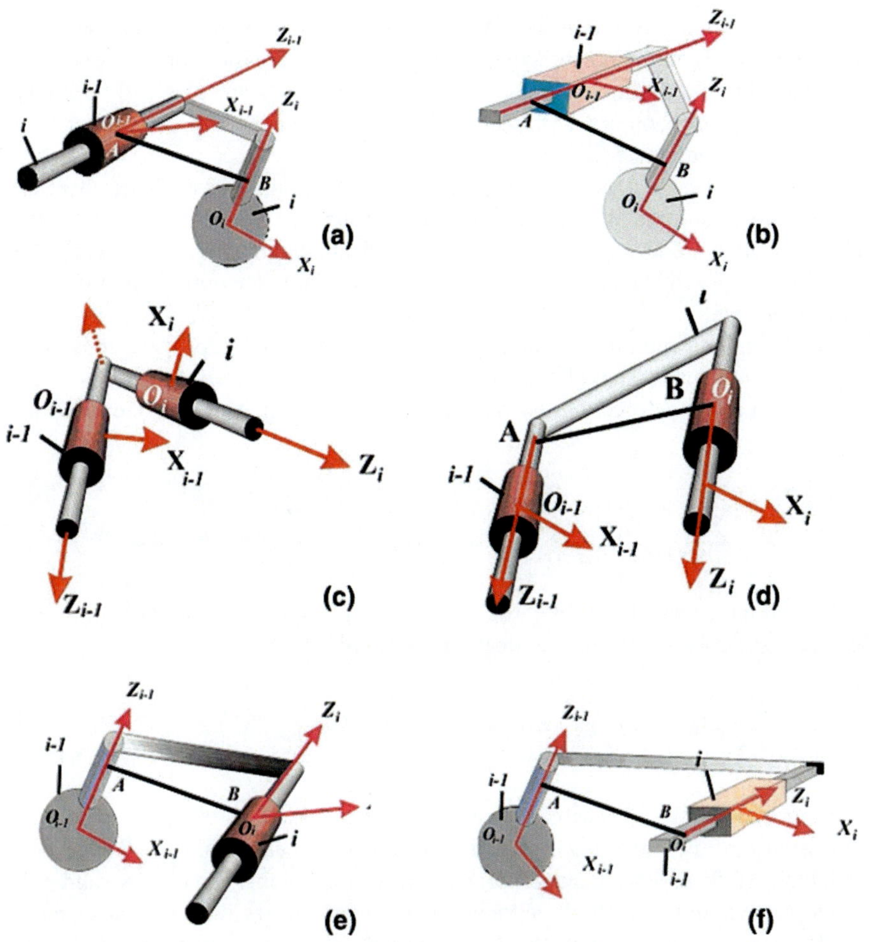

Fig. 3.4 Examples of relative position of joints with various zero parameters

is possible in case, when the transformation parameters are dependent or equal to zero. Cases, when the transformation parameters are equal to zero due to the mutual orientation of the coordinate axes, are discussed in Sect. 3.4. Let us consider the possible cases, when the transformation parameters are dependent. Let, in the coordinate systems being transferred, the axes $O_{i-1}Z_{i-1}$ and O_iZ_i are parallel (Fig. 3.4d), then the transformation parameters d_i and b_i, and θ_i, β_i characterize the linear and angular displacements in the same direction. As a result, these expressions for the works in Eq. (3.4) appear to be similar and as a result of bringing the similar members, the number of summands is reduced to four. Thus, in this case the repeated parameters should be assigned a value of zero ($b_i = \beta_i = 0$), which is equivalent to a reduction of summands in the expression (3.4). There is another case of the transformation, when there are dependent links. For example, if during the transformation, according to some two parameters, the system is moved relative to an axis twice. Then in Eq. (3.3) one summant will decrease, because expression for the work of the generalized forces during the displacement relative to the same axis will be similar terms. In this case, both transformation parameters are dependent, and the repeated transformation parameter is taken as zero.

In some cases, such as, for example, in extra-constraint mechanisms, there are repeated kinematic chains, which allow obtaining additional equations for the works. These equations reduce the number of independent equations (3.4), i.e., each new equation composed in addition to the system (3.4) reduces the dimensionality of the space constraints.

To obtain the conditions for determining the number of independent constraints of a closed kinematic chain, let us in Eq. (3.1) totalize basic works of similar forces and pairs of reaction forces (let us totalize vertically). Then, we obtain the following equation:

$$\sum_{k=1}^{n} F_k^1 \delta d_k + \sum_{k=1}^{n} M_k^1 \delta \theta_k + \sum_{k=1}^{n} F_k^2 \delta a_k + \sum_{k=1}^{n} M_k^2 \delta \alpha_k + \sum_{k=1}^{n} F_k^3 \delta b_k + \sum_{k=1}^{n} M_k^3 \delta \beta_k = 0.$$

To find out in which case a particular summant in this equation turns into zero, let us consider separately any summant, for example,

$$\sum_{k=1}^{n} F_k^1 \delta d_k = F_1^1 \delta d_1 + F_2^1 \delta d_2 + \cdots + F_{n-1}^1 \delta d_{n-1} + F_0^1 \delta d_0 = 0.$$

Here, the reaction forces cannot equal to zero reasoning from the conditions of existence of the joint. Transformation parameters in each joint are independent variables. All these leads to an important conclusion that this equation is valid only under the condition that all of the transformation parameters having the same name (in this case, it is d_k) amount to zero. Therefore, if some number—$S < 6$ of zero transformation parameters having the same name—are available in all equations of the system (3.4), then the rank of the equations coefficient matrix (3.4) and, respectively, the dimensionality of the space decreases for the number of S. In this case, the maximum dimensionality of the space described by the system (3.4) is equal to $R = 6 - S$.

Therefore, number S can be determined as the least number of zero parameters having the same name in each of the equations in the system. For example, if some mechanism in the system of six equations of the form (3.4) is equal to zero, in the first equation, three parameters, in the second—four, in the third—four, in the fifth—three, and in the sixth—two parameters, then $S = 2$, because in all equations of the system, at least two parameters are equal to zero. In this case, for this mechanism, the dimensionality of the space is equal to $R = 6 - S = 4$, i.e., the number of independent parameters defining the constraints is 4. It follows that the number of dependent constraints S is equal to the least number of zero parameters having the same name and available in all Eq. (3.4) of the system. Analysis of the matrix of the system coefficients (3.4) allows determining the number of dependent constraints. From a practical point of view, to determine S for each closed loop of the mechanism or manipulator, it is enough to make a table (matrix) of the transformation parameters. Then using the table, the smallest zero parameter is defined among those having the same name in all rows. This number defines the number of dependent constraints S.

3.7 Derivation of the Equation for Determining the Degree-of-Freedom Considering Dependent Constraints

To determine the degree-of-freedom of a one-loop mechanism, let us use a known method of open kinematic chain. Given an open kinematic chain, degree-of-freedom is equal to M—the sum of the degrees-of-freedom of joints. On some link of this kinematic chain, six constraints can be imposed to secure this link. However, among the available constraints, there are S dependent constraints. Therefore, in order not to change the dimensionality of the space, the locking element should have $(6 - S)$ independent constraints. Now, some minimum number of $S < 6$ dependent constraints present in all joints. In this case, the number of independent constraints becomes equal to $(6 - S)$. Therefore, a one-loop mechanism is formed, in which the total number of degrees-of-freedom (M) reduces for a number equal to $(6 - S)$. For a one-loop mechanism, degree-of-freedom can be obtained in the form of

$$W = M - 6 + S. \tag{3.5}$$

It should be noted that the last link can be also imposed with six constraints, while the number of dependent constraints in the kinematic chain becomes zero ($S = 0$). However, in this case, to get a mechanism with a degree-of-freedom equal to $W = 1$, it is necessary to increase the number of degrees-of-freedom and the number of links. Thus, for example, to obtain a mechanism with a degree-of-freedom $W = 1$ with $S = 0$, according to Formula (3.5), the mechanism must have seven degrees-of-freedom, i.e., six links with one-degree-of-freedom joint.

Dependence similar to (3.5) was obtained earlier in the study [34] and was given also in the study [40]. However, these and other studies did not give a method for determining the number of dependent constraints applicable to any mechanism and parallel manipulator.

The proposed method for determining the number of dependent constraints is based on the transformation of six parameters of coordinate systems related to the links. At the same time, zero transformation parameters are set. Then using the table (matrix) containing zero parameters, the number of dependent constraints of a closed kinematic chain is determined.

Degree-of-freedom of a multiloop mechanism can be obtained depending on

$$W = M - 6k + \sum_{i=1}^{k} S_1, \quad (3.6)$$

where k is the number of independent loops and S_i is the number of dependent constraints of the i-th loop.

3.8 Sequence of the Mechanism Design Analysis Using the Zero Parameters

The sequence of actions performed during the mechanism design analysis is as follows:

1. Number the moving links. It is desirable to mark the fixed supports with the digit 0.
2. Determine the number of independent loops.
3. With each of the links of the selected loop, including the fixed link, connect the coordinate system in accordance with Sect. 3.2.
4. In each loop, successively combining the coordinate system, determine the transformation parameters and make a table of transformation parameters having excluded dependent parameters. Then determine the rows, in which the minimum number of parameters is equal to zero. This minimum number of zero parameters is equal to the number of dependent constraints of the loop being analyzed.
5. According to Formula (3.5) or (3.6), determine the degree-of-freedom of the mechanism.

Below, using the above sequence, the mechanism design is analyzed, for which there are discrepancies between the actual and estimated (by conventional means) values of the NDF.

3.9 Examples of the Mechanism and Manipulator Design Analysis

In Sect. 3.1, it was shown that for a closed mechanism (Fig. 3.1b) formed by fixing the last link of the telescopic mechanism, the real degree-of-freedom does not coincide with the degree-of-freedom found according to one of the traditional structural formulas. However, use of the method obtained in this study and dependency (3.6) allows determining the actual degree-of-freedom. To verify this, let us analyze the design of the telescopic mechanism according to the zero-parameter method in the sequence described in Sect. 3.7. As a result, the following is obtained:

1. In the telescopic mechanism (Fig. 3.1b), there are two moving links (1, 2) and a fixed link denoted by the digit 0.

Table 3.2 Transformation parameters for telescopic joint

Links forming joints	Transformation parameters					
	d_i	θ_i	a_i	α_i	b_i	β_i
0–1	d_1	0	0	0	0	0
1–2	d_2	0	0	0	0	0
2–0	d_3	0	0	0	0	0

2. The mechanism has one loop.
3. All links related to the axis Z_0 and Z_i ($i = 1, 2$) are directed along the axes of the kinematic pairs, i.e., they coincide. Related to the links, axes X are directed perpendicular to the axis Z and are located in one plane.
4. As a result of coordinate system transformations, the transformation table (Table 3.2) was built. Analysis of the table shows that the number of dependent constraints is $S = 5, M = 3$.
5. According to Formula (3.5), NDF of the mechanism being considered is $W = 2$.

Thus, using a simple example of a closed telescopic mechanism, it was determined that the value of the DOF obtained by the zero-parameter method coincides with the real DOF.

Below using the zero-parameter method, NDF is determined for three spatial articulated four-link lever mechanisms with rotational kinematic pairs. For these mechanisms, the traditional formula of Malyshev-Somov gives the following DOF value:

$$W = 6n - \sum_{i}^{5} iC_i = 6 \times 3 - 5 \times 4 = -2.$$

However, this NDF value does not correspond to the actual NDF values given for the articulation four-link mechanisms differing from each other by the design topology. Further, it is shown that it is the zero-parameter method that allows determining the actual NDF values considering the mechanism design topology.

First, the mechanism (Fig. 3.5) is discussed, in which the axes of the rotational pairs are mutually parallel. By analyzing the mechanism by the zero-parameter method in the sequence given in Sect. 3.7, it is found that

1. The mechanism has three moving links 1–3 and a support.
2. The mechanism has one loop.
3. In accordance with the recommendations, coordinate axes are chosen, while the parallel axes Z_0, Z_1, Z_2, and Z_3 *are directed* along the axes of rotational kinematic pairs. Axis X_0 is selected correctly. Axes X_1, X_2, X_3 are directed perpendicular to the axes, as shown in Fig. 3.5. The choice of the center of the system of coordinates O_0, O_1, O_2, O_3, in this example, is not critical, as the task of the analysis is to determine the parameters equal to zero, but not the values of the transformation parameters.

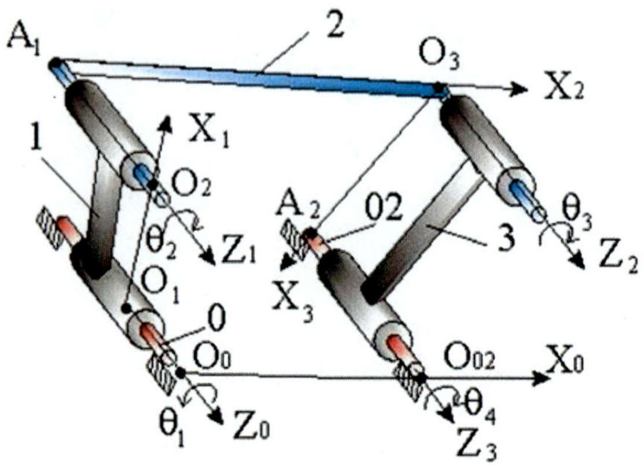

Fig. 3.5 Articulation mechanism with parallel axes

4. Transformation of the coordinate systems is performed in the sequence given in Sect. 3.3 and Table 3.3 is made, while it was taken into account that due to the parallelism of the axes Z, displacements $b_i = 0$ and $\beta_i = 0$. In this case, axes Z_{i-1} *and* Z_i are parallel, so the displacements and rotations are enough to produce once—around the axis Z_{i-1}. Also, due to the fact that axes Z_{i-1} *and* Z_i are parallel, it is not necessary to turn around the axis X_{i-1}, i.e., $\alpha_i = 0$. The analysis of the transformation parameters shows the variable parameters 4 ($\theta_i =$ Var), i.e., $M = 4$. The table shows that in the rows the minimum number of zero parameter is three, i.e., $S = 3$.
5. According to Formula (3.5), NDF $W = 1$ is determined.

This value corresponds to the actual NDF of the articulated four-link mechanism with parallel axes of the rotational pairs. Let us now consider the four-link level mechanism (Fig. 3.6), called a spherical mechanism, with rotational pairs, in which rotation axes intersect in one point O. In Fig. 3.6, links 0–3 are denoted, and axes Z_i related to links are selected; these axes are directed along the axes of rotational kinematic pairs. Direction of the axes X_0 is selected perpendicular to the plane Z_0OZ_3. Other axes X_i are perpendicular to the plane, where axes Z_{i-1} and Z_i are located.

To align the coordinate system $O_0X_0Z_0$ with the coordinate system $O_1X_1Z_1$, let us perform the following displacements: displace the system $O_0X_0Z_0$ along the axes O_0Z_0 for the value of $d_1 = O_0O$; turn the axes OX_0 for the angle of θ_1 till it coincides

Table 3.3 Transformation parameters for the mechanism with parallel axes

Links forming joints	Transformation parameters				
	d_i	θ_i	a_i	α_i	b_i	β_i
0–1	$d_1 = O_0O_1$	θ_1	$a_1 = O_1O_2$	0	0	0
1–2	$d_2 = O_2A_1$	θ_2	$a_2 = A_1O_3$	0	0	0
2–3	$d_3 = 0$	θ_3	$a_3 = O_3A_2$	0	0	0
3–0	$d_4 = A_2O_{02}$	θ_4	$a_4 = O_0O_{02}$	0	0	0

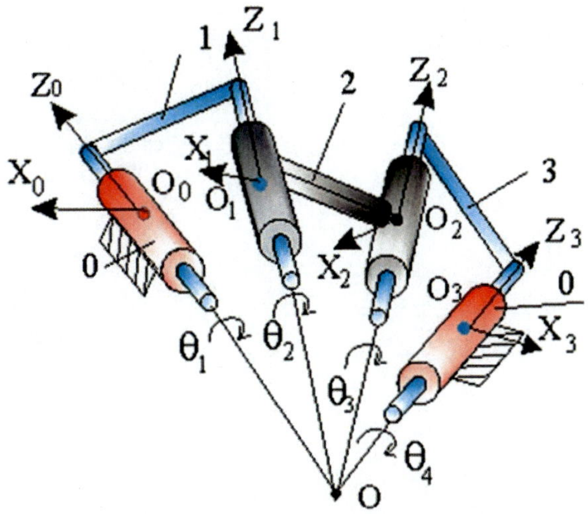

Fig. 3.6 Four-link spherical mechanism

with the direction of the axis O_1X_1; shift along the axis O_1X_1 is not done, therefore, $a_1 = 0$; turn the system from the obtained position for the angle of $\angle\alpha_1 = \angle O_0OO_1$; displace the obtained system along the axis OZ_1 for the value of $b_1 = OO_1$; the turn around the axis O_1Z_1 is not needed, so let us suppose that $\beta_1 = 0$. Similar actions are performed to identify transformation parameters in other kinematic pairs. As a result of the transformation, dependent parameters are determined, namely, $b_1 = d_2$; $b_2 = d_3$; $b_3 = d_4$; $b_4 = d_1$. In accordance with the findings obtained in part 5, let us suppose that $b_1 = b_2 = b_3 = b_4 = 0$. The obtained transformation parameters for the kinematic pairs with the elimination of dependent parameters are summarized in Table 3.4. The table analysis shows that $M = 4$, i.e., $\theta_i = \text{Var}$, $S = 3$. Therefore, according to Formula (3.5) the degree-of-freedom of the mechanism is $W = 1$.

Example 1 As an example that also discloses the possibilities of the described zero-parameter method, let us analyze the design of the Bennett mechanism (Fig. 3.7). Following the accepted above (3.6) sequence, it is established that

1. The mechanism consists of four moving links, and let us number them. The fixed supports are denoted with 0.

Table 3.4 Transformation parameters for the spherical mechanism

Links forming joints	Transformation parameters					
	d_i	θ_i	a_i	α_i	b_i	β_i
0–1	$d_1 = O_0O$	θ_1	0	α_1	0	0
1–2	$d_2 = OO_1$	θ_2	0	α_2	0	0
2–3	$d_3 = OO_2$	θ_3	0	α_3	0	0
3–0	$d_4 = OO_3$	θ_4	0	α_4	0	0

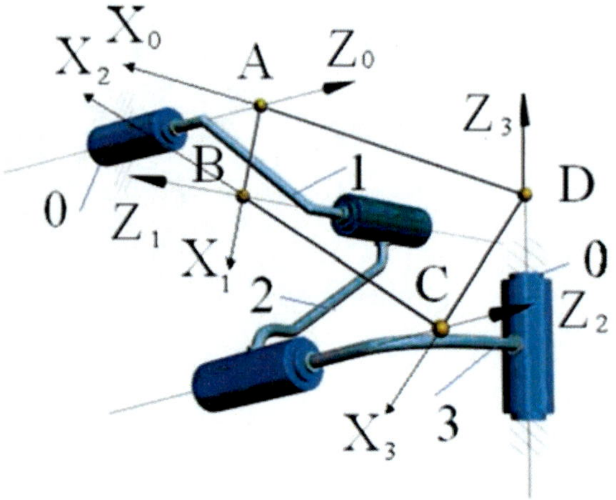

Fig. 3.7 Bennett's mechanism diagram

2. It is obvious that the mechanism is one loop.
3. Coordinate axis associated with the links are selected according to the above recommendations. Fixed axis Z_0 is drawn over the axis of the joint formed by the link 0–1. Axes Z_{i-1} ($i = 2, \ldots, n$) will be directed over the axis of joints made of the links $i-1$, i. It should be noted that an important feature of the Bennett mechanism is that the common base of the perpendiculars to the adjacent axes coincide. Therefore, in the points *A*, *B*, *C*, and *D*, perpendiculars to the adjacent axes intersect. For example, at the point *A*, the perpendicular *AD* to the axis Z_0 and the perpendicular *AB* to the axis Z_1 *intersect.* Using this feature, over the line *AB* axes X_1 *is directed.* In this case, $X_1 \perp Z_0, Z_1$. Similarly, over the line *BC*, the axis X_2 ($X_2 \perp Z_1, Z_2$) will be directed over the straight line *CD*—axis X_3 ($X_3 \perp Z_2, Z_3$). The axis X_0 is directed over the line *AD* perpendicularly to the axes Z_0 and Z_3. After selecting all the coordinate systems related to the links, the previous system of coordinates is successively aligned with the following starting from $O_0X_0Z_0$, while traversing along the mechanism loop.

To return to the original position, transformation is performed from the coordinate system $O_3X_3Z_3$ to the system $O_0X_0Z_0$. Thus, the transformation parameters of the coordinate system are obtained. Characteristics for each articulation transformation parameter are entered in Table 3.5. Analysis of the data in Table 3.5 shows that in all joints parameters are $d_i = b_i = \beta_i = 0$. Therefore, $S = 3$.

4. Using Formula (3.6), it is found that the NDF of the Bennett mechanism is equal to one, as $W = 4 - 6 + 3 = 1$, where $M = 4$.

This result coincides with the true value of the degree-of-freedom of the Bennett mechanism. It should be noted that none of the previously known methods of calculating the DOF does not allow determining the actual DOF of the Bennett mechanism.

Table 3.5 Transformation parameters for the Bennett mechanism

Links forming joints	Transformation parameters					
	d_i	θ_i	a_i	α_i	b_i	β_i
0–1	0	θ_1	AB	α_1	0	0
1–2	0	θ_2	BC	α_2	0	0
2–3	0	θ_3	CD	α_3	0	0
3–0	0	θ_4	AD	α_4	0	0

Example 2 To confirm the universality of the zero-parameter method, let us determine the DOF of the parallel manipulator described by *Jing-Shan Zhao* et al. in their study [18] at page 622 according to the accepted sequence in Sect. 3.6.

1. Let us number the moving links from 1 to 25 (Fig. 3.8a). The fixed supports are denoted with 01–04 (In fact, they represent one fixed link 0). Joints of the link 7 are denoted with 71–74.
2. To determine the number of closed loops, let us build a topological space presented in Fig. 3.8b, having denoted the links with the graph nodes, and joints with edges. While the link 0 will be represented with four vertices 01–04, and the link 7 with the vertices 71–74, edges connecting these vertices denote the transformation from one coordinate system related to the link, to another, e.g., from system 01 to the coordinate system 02. Let us use the known Euler formula:

$$P - L + k = 2, \quad (3.7)$$

where P is the number of vertices; L is the number of edges; and K is the number of faces.

Faces are the loops to be found. Based on the structure, o$P = 32$, $L = 36$. According to Formula (3.7), the number of loops $K = 6$ is found.

3. Selection of the coordinate axes in more detail is illustrated in the example of the loop as shown in Fig. 3.8c. Axes Z_i and X_i are directed according to the rules specified in Sect. 3.1. Randomly selected axes X_{01} and X_{02} are directed similarly.

Coordinate dihedral moves along the loop $P_1B_1B_2P_2P_1$ from the position $Z_{01}X_{01}$ through the links 1–2–3–4–5–6–71–72–8–9–10–11–12–13–02 and returns into its starting position. Identified transformation parameters are summarized in Table 3.3. In this case, variable parameters are denoted with q. Constant parameters that are the same everywhere are denoted by the letter c. We do not specify certain values q and c, as their values are not significant and do not affect the result of the problem. Comparative analysis of the zeros in the rows of Table 3.6 demonstrates that the minimum number of zero parameters in the rows is three. Hence, in this loop $S = 3$. Repeat the similar actions for the next loop, moving the coordinate dihedral $Z_{01}X_{01}$ through the links 1–2–3–4–51–54–14–15–16–17–04 and returning it to the starting

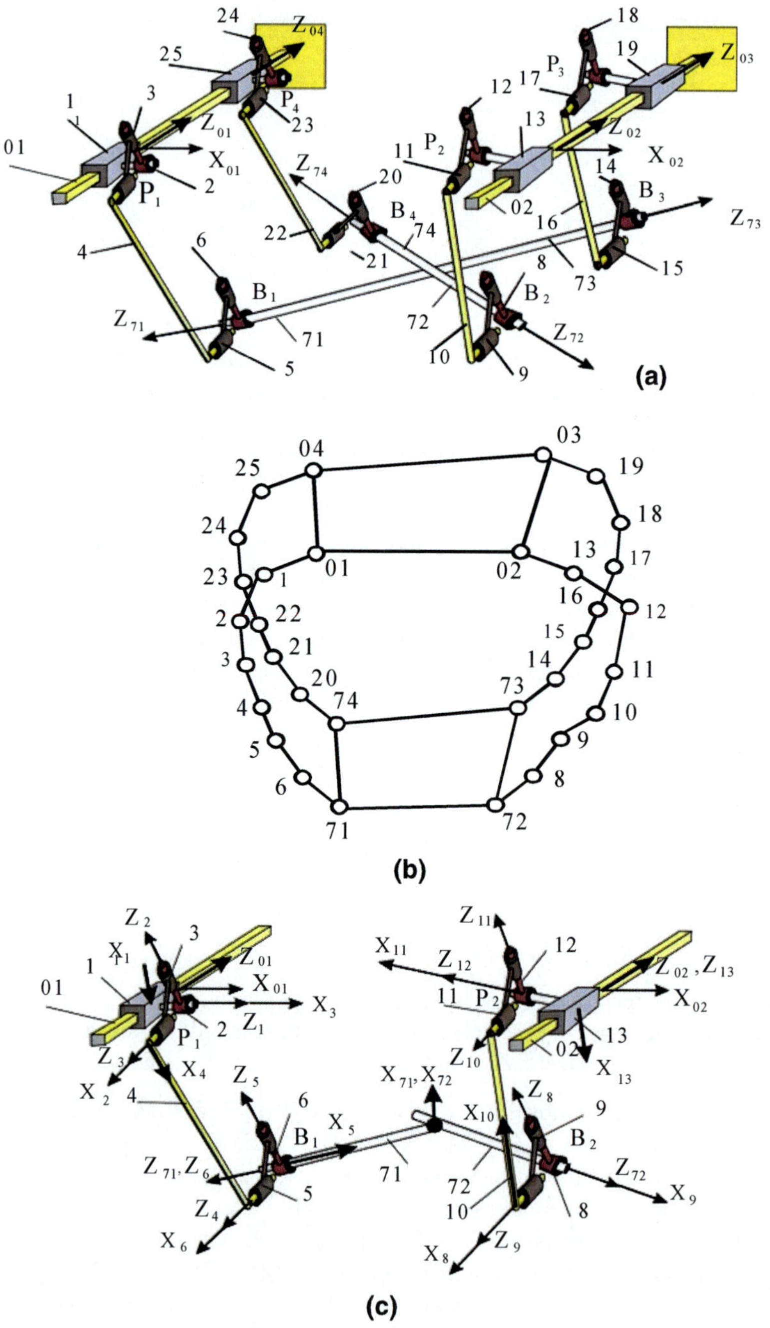

Fig. 3.8 Parallel manipulator

Table 3.6 Transformation parameters for the manipulator

Links forming joints	Transformation parameters					
	d_i	θ_i	a_i	α_i	b_i	β_i
01–1	q	c	0	c	0	0
1–2	c	q	0	c	0	0
2–3	0	q	0	c	0	0
3–4	c	q	c	0	0	0
4–5	c	q	0	c	0	0
5–6	c	q	0	c	0	0
6–71	c	q	0	c	0	0
71–72	c	0	0	c	0	0
72–8	c	q	0	c	0	0
8–9	0	q	−0	c	0	0
9–10	c	q	c	0	0	0
10–11	c	q	0	c	0	0
11–12	0	q	0	c	0	0
12–13	c	q	0	c	0	0
13–02	q	c	0	0	0	0
02–01	0	0	c	0	0	0

position. Then, move the coordinate dihedral $Z_{04}X_{04}$ through the links 17–16–15–14–54–53–10–11–12–13–03–04. Similarly, let us move the coordinate dihedral $Z_{03}X_{03}$ through the links 13–12–11–10–53–52–6–7–8–9–02–03. Constructed based on these results tables of transformation matrix, elements comprise rows with the minimum number of zero parameters equal to three.

By moving coordinate dihedral $Z_{01}X_{01}$ along the loop consisting of the links 02–03–04–01, and based on the results of the parameter table analysis, the minimum number of zero parameters in rows equal to 4 is obtained.

By moving the coordinate dihedral along the path of 51–52–53–54–51 and having built the table of transformation parameters, it is seen that the minimal number of zero parameters in this loop is equal to 4.

Thus, in four lateral loops $S = 3$, and in the bottom and top loops $S = 4$.

4. According to Formula (3.6), let us determine the NDF.
5. $W = 20 - 6 \times 6 + 4 \times 3 + 2 \times 4 = 4$.

The indicated NDF value coincides with the value W obtained in the study [40].

3.10 Construction of Rational Mechanisms

One of the synthesis problems is to eliminate dependent constraints and construction of rational mechanisms. If during the analysis of the mechanisms, it is found that $S > 0$, it means that the mechanism has dependent constraints. To eliminate these constraints, it is necessary for the transformation parameters' table to amend the rows with zero elements, having replaced the zero element in the selected row with a variable transformation parameter. This problem is a multi-variant one. However, the number of variants reduces if we keep in mind that the joints with the DOF exceeding three are non-constructive, or that the input and output links with the support usually form one-DOF joints. The designer in each case decides in which joints it is needed to change mobility and which types of mobility to add. The number of additionally introduced degrees-of-freedom is determined according to Formula (3.5) based on that as a result to obtain the given NDF ($W = 1$) with $S = 0$. For example, let us eliminate the dependent constraints in the above Bennett mechanism (Fig. 3.7). Using Formula (3.5), it can be found that $M = 6 + W = 7$, i.e., it is necessary to the available four variable transformation parameters to add three variable parameters.

One solution option for the problem is to replace the rotational joints 1–2 and 2–3 with the cylindrical and spherical kinematic pair, respectively. As a result, the structure of the rational four-link mechanism is obtained as is illustrated in Fig. 3.9.

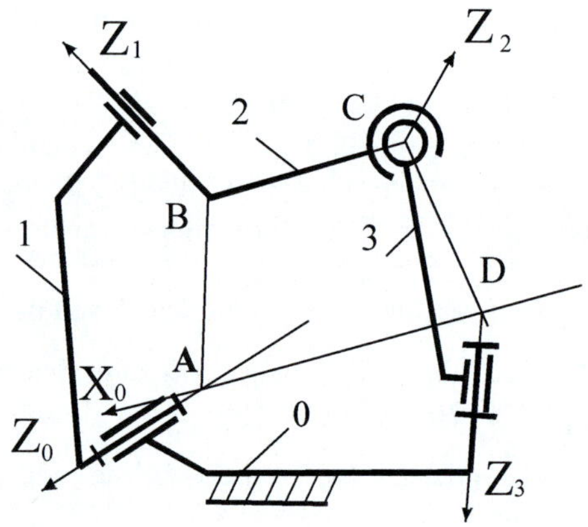

Fig. 3.9 Mechanism without dependent constraints

Chapter 4
Analytical Description of the Links of Spatial Mechanisms and Manipulators by Redundant Parameters Method

Abstract The chapter offers the analytical method for describing link positions in mechanisms and manipulators that is applicable to any kinematic pairs and that provides simple final matrix expressions, and one allowing for expanding application area of the existing matrix methods. This method is based on the use of 8 transformation parameters of coordinate systems. Examples of application are given, and versatility and advantages of the redundant parameters are justified.

Keywords Matrix transformation methods · Analytical determination of link position · 8 Parameters

4.1 Construction of Mathematical Models of Mechanisms

Mathematical models of multilink spatial mechanisms and parallel manipulators with kinematic chains containing closed loops should possess the greatest possible versatility. These models should allow producing certain results using PC in solving problems of analysis and synthesis of the design, kinematics, and dynamics, as well as control problems.

Construction of mathematical models of the mechanisms and manipulators, as a rule, begins with the selection of the research apparatus for the analytical description of the links position. It is known [1–3, 44] that the description of the mechanisms and manipulators link positions uses analytical apparatus, based on the methods of screws quaternion algebra, matrix method, etc. Many scientific publications as well as engineering practice evidence that the representation of Denavit-Hartenberg (D-H representation) [29] using a special choice of the axes of the coordinate system and homogeneous dimensionality matrices 4×4 to describe the transformation of the coordinate systems are the most widely used in the research of the manipulators.

Links and kinematic pairs (KP), as is known, are characterized by geometric and kinematic parameters. The geometrical parameters represent the dimensions of the links, the angles of orientation of the axes, planes, etc. These parameters in the movement of the mechanism links do not change. Kinematic parameters represent variable angular and linear values, the number of which determines the DOF

K. S. Sholanov, *Parallel Manipulators of Robots*, Mechanisms and Machine Science 92,
https://doi.org/10.1007/978-3-030-56073-7_4

of the kinematic pair. When applying matrix methods, usually links are related to the coordinate systems. By transforming the coordinate systems, expressions defining links position relative to each other and relative to the fixed coordinate system are obtained. Elements of transformation matrices or their arguments when transformation elements represent functions are called transformation parameters.

In the simulation of the position by matrix methods, geometric and kinematic parameters of the links and KP are used as transformation parameters. For example, in D-H representation, four parameters of the links and KP are also transformation parameters. Given four parameters limit the scope of application of D-H representation, since they can describe the position of the links of mechanisms and manipulators having only rotational, translational, and sometimes cylindrical KP, axes of which are parallel or intersect at right angles. In order to expand the possibilities of the D-H representation and to cover all kinds of KP, the study [42] uses six transformation parameters. However, in this case, the number of coordinate system transformations equals the double number of kinematic pairs, which leads to complication of the final matrix expression.

Here the task to obtaining an analytical method of describing the position of the links of mechanisms and manipulators, suitable for all KPs provide a simple final matrix expression. Thus, it is proposed to extend the scope of application of the matrix methods.

4.2 Selection of the Coordinate Systems for the Transformation by eight Parameters

Related to the links systems of coordinates i–1 and i can have, in general case, intersecting axes Z_{i-1} and Z_i. These axes have a single common perpendicular. If the i–1-coordinate system is moved through this common perpendicular till it fully coincides with the i system of coordinate, this transformation will be described by a transformation matrix with six transformation parameters. However, to describe the position of the links of manipulators and mechanisms with complex spatial configuration and topology, such combination in most cases is not acceptable. The reason is that when the position of the axes Z_{i-1} and Z_i is changed, the length and position of the common perpendicular change too. Moreover, if the common perpendicular in not the link parameter or KP, the transform matrix would comprise useless values that are not link parameters and KP.

The problem posed here is solved by transforming the i-coordinate system into i–1 system not via the common perpendicular as, for example, in the study [29], but via obviously selected (any) line crossing the axes Z_{i-1} and Z_i [43, 45]. This transformation is illustrated by the kinematic chains shown in Fig. 4.1. It shows the links k and k–1 connected by two-DOF cylindrical kinematic pair with kinematic parameters of $AB = a_i$ и α_i and axes X and Z of the system of coordinates with numbers i and i–1, while the axis Z_{i-1} is directed along the axis of the cylindrical

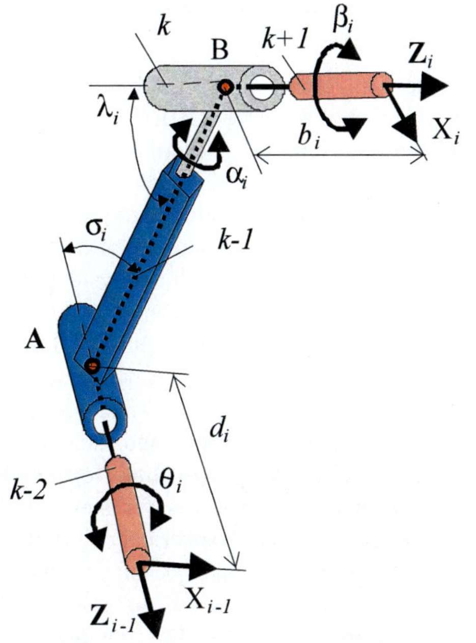

Fig. 4.1 Selection of the coordinate system

kinematic pair formed by the link k–1, and the link k–2 that is related to the system i–1. This KP is characterized by the kinematic parameters d_i and θ_i. Axes Z_{i-1} forms the angle σ_i with the line AB.

The axes Z_i is directed along another cylindrical pair (with kinematic parameters b_i and β_i) from the link k and link $k + 1$ related to the system of coordinates i. This axis forms an angle λ_i with the straight-line AB. Kinematic chain involved in the transformation is characterized by geometrical parameters (σ_ι, λ_ι) and variable kinematic parameters d_i, θ_i, a_i, α_i, b_i and β_i. It should be noted that a number of the abovementioned kinematic parameters in each case may become the geometric parameters and vice versa.

Hereinafter, the line AB passing through the link and intersecting the axes Z_{i-1} *and* Z_i, *is referred to as direct connecting link distance.*

Thus, to include all of the link parameters and KP in the row of transformation parameters and cover several KP with one transformation, the coordinate system O_{i-1} X_{i-1} Y_{i-1} Z_{i-1} is moved through the connecting line AB until it fully coincides with the coordinate system O_i X_i Y_i Z_i. As it is shown below, this requires a combination of eight consecutive stages (steps) of transformation of the coordinate systems O_{i-1} X_{i-1} Y_{i-1} Z_{i-1}, i.e., introduction of eight transformation parameters.

Body position, as is known, is determined by six independent values. The proposed method uses redundant parameters, where some of the transformation parameters are dependent.

4.3 Peculiarities of the Selection of the Coordinate Systems

Success in solving the set problem, as well as the amount of labor spent for the preparation of the mathematical model depends largely on the proper selection and skills in choosing the coordinate system and connecting line. Moreover, the coordinate system as in the example in Fig. 4.1 does not necessarily have to be related directly to the links forming the kinematic pairs. Moreover, the coordinate systems can be numbered independently of the number of links in the manipulator mechanism. Selection of the coordinate system and the connecting line must satisfy specific requirements, namely, all geometric and kinematic parameters, and KP of the links must be transformation parameters of the coordinate systems.

To satisfy this requirement, while selecting the coordinate axes, the concept of characteristic elements of the kinematic pairs (CEKP) is introduced. As the CEKP, a point or a line is accepted, and such point or line must have the same speeds and belong simultaneously to the links forming KP. In some cases, CEKP is represented by the characteristic surfaces, for example, surface of teeth, cam, surfaces of the slot for the pin in two-DOF spherical pair with a pin, etc.

Thus, for the rotational, prismatic, and cylindrical kinematic pair, a characteristic element is a straight line, for the spherical pair, the characteristic element is a point.

Recommendations for choosing the coordinate system are given below.

- The construction of the coordinate axes related to the supports is given certain freedom. (Supports and coordinate systems related to the supports are usually denoted with 0, 01, 02, etc.) Axes Z_0 is directed by characteristic elements of the KP formed by the support and the moving link. Center of coordinates coincides with CEKP if the characteristic element of the KP is a point. Axes X_0 related to the support is directed through the point of intersection of the axes Z_0 and Z_1, when axes intersect perpendicular to the plane, where the axes Z_0 and Z_1 are located, or along the link toward the axis Z_1 of the conjugated link. Axis Y_0 is directed to form the appropriate coordinate system. If a spheric pair with a pin is the support then the axis Y_0 is directed along the normal to the plane of the slot.
- Axes Z_{i-1} and Z_i are directed along the CEKP.
- Axis X_{i-1} ($i = 2, \ldots, N + 1$) is directed through the point of the intersection of axes Z_{i-1} and Z_i, if the axes Z intersect along the perpendicular to the axes Z_{i-1} and Z_i or toward the axis Z_i (in the general case).
- The axis Y_{i-1} is constructed so that the obtained coordinate system O_{i-1} X_{i-1} Y_{i-1} Z_{i-1} remains on the right side. If KP is a two-DOF spherical pair, then the axis Y_{i-1} is directed along the normal to the plane of the slot.

After selecting the first coordinate system O_{i-1} X_{i-1} Y_{i-1} Z_{i-1}, the connecting line *AB* is selected. At the same time, it should be noted that through the point *A* of the connecting line axis, Z_{i-1} passes; this point can be the point of intersection of 2–3 rotational kinematic pairs. Straight-line *AB* passes through the link or through the cylindrical KP. Point *B* of the line *AB*, similarly to the point *A* is the point of intersection of the line with the axis Z_i and the point of intersection of 2–3 rotational kinematic pairs. Axis Z_i passes through the point B and is directed along the CEKP in the kinematic chain which is the last covered with the transformation. The position of the axis Z_i will be used as the starting position for further transformation as given below.

Signs of the correctly chosen coordinate systems is that

- combining the coordinate systems is performed using a minimum number of the transformation parameters, when the maximum number of parameters is zero;
- when the position of the links varies, only variable transformation parameters called articulation parameters are changed;
- when the change of the link position does not change the transformation parameters characterizing the geometrical values (dimensions, twist angles) called link parameters.

4.4 Sequence of the Coordinate Systems' Transformation

Let us consider the sequence of transformations using redundant parameters [23, 24]. Suppose that the coordinate system $O_{i-1}\ X_{i-1}\ Y_{i-1}\ Z_{i-1}$ and $O_i\ X_i\ Y_i\ Z_i$ (Fig. 4.2) related to the links are given. It is required to transform the i-coordinate system into i–1 system, and through the line AB, determine the homogeneous transformation matrix

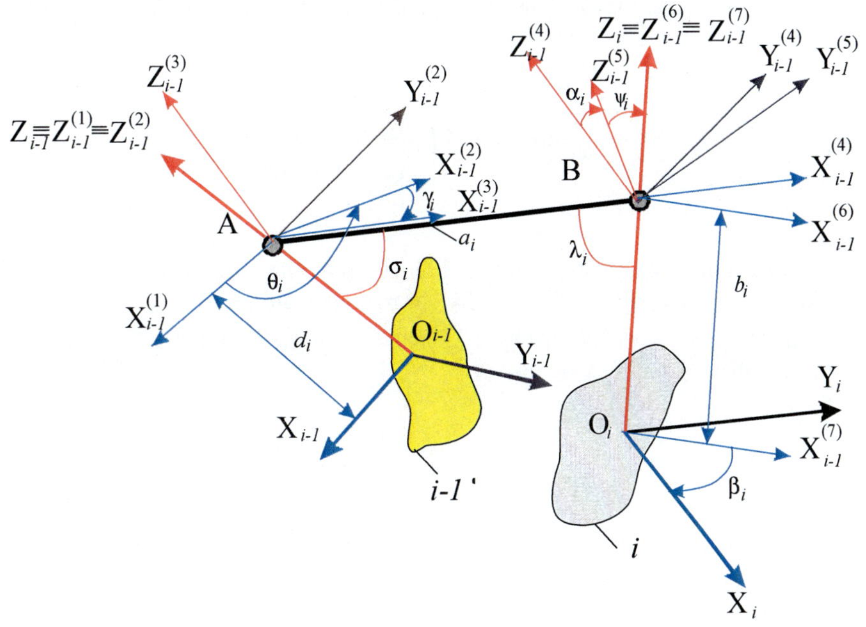

Fig. 4.2 Sequence of movements

$\mathbf{A}_i^{i-1}$. This matrix plays a key role in determining the position of the links relative to each other and the fixed coordinate system. Transformation matrix elements $\mathbf{A}_i^{i-1}$ are determined by combining the *i*–1 system with the *i*-coordinate system by successive angular and linear displacements.

Below is the sequence of movements of the *i*–1-coordinate system till its full alignment with the *i*-coordinate system. These movements are carried out as indicated above through the line *AB* so that all geometric and kinematic parameters of links and KP are included as transformation parameters in the matrix $\mathbf{A}_i^{i-1}$.

1. The *i*–1 system is shifted along the axis Z_{i-1} for the distance of d_i, to align the center of the coordinate system $O_{i-1}\ X_{i-1}\ Y_{i-1}\ Z_{i-1}$ with the point *A*. While the axis X_{i-1} takes the position $X_{i-1}^{(1)}$. *Here and in Fig. 4.2 upper index indicates the number of intermediate position of the coordinate system.* This movement is described by the transposition matrix

$$\boldsymbol{T}(Z_{i-1}, d_i) = \begin{pmatrix} 1\,0\,0\,0 \\ 0\,1\,0\,0 \\ 0\,0\,1\,d_i \\ 0\,0\,0\,1 \end{pmatrix}.$$

2. Then the axis $X_{i-1}^{(1)}$ is rotated for the angle of θ_i around the axis $Z_{i-1}^{(1)}$, so that the new axis $X_{i-1}^{(2)}$ is located in the plane passing through the axis $Z_{i-1}^{(1)}$ and the line *AB*, while the axis $Y_{i-1}^{(2)}$ becomes perpendicular to the line *AB*. The rotation is described by the rotation matrix

$$\boldsymbol{R}(Z_{i-1}^{(1)}, \theta_i) = \begin{pmatrix} C\theta_i & -S\theta_i & 0\,0 \\ S\theta_i & C\theta_i & 0\,0 \\ 0 & 0 & 1\,0 \\ 0 & 0 & 0\,1 \end{pmatrix}.$$

Here the following notations are used for the functions: *Sin—S*, *Cos—C*.

3. The coordinate system $AX_{i-1}^{(2)}Y_{i-1}^{(2)}Z_{i-1}^{(2)}$ is rotated relative to the axis $Y_{i-1}^{(2)}$ for the angle of $\gamma_i = \frac{\pi}{2} - \sigma_i$ until the axes $X_{i-1}^{(3)}$ and the line *AB* coincides. This rotation is described by the rotation matrix

$$\boldsymbol{R}\left(Y_{i-1}^{(2)}, g_i\right) = \begin{pmatrix} C\gamma_i & 0 & S\gamma_i & 0 \\ 0 & 1 & 0 & 0 \\ -S\gamma_i & 0 & C\gamma_i & 0 \\ 0 & 0 & 0 & 1 \end{pmatrix}.$$

4. Now the shift is performed along the axis $X_{i-1}^{(3)}$ for the distance of α_i till the coordinate system $AX_{i-1}^{(3)}Y_{i-1}^{(3)}Z_{i-1}^{(3)}$ coincides with the point *B*. This shift is described by the transposition matrix

$$T\left(X_{i-1}^{(3)},\alpha_i\right)=\begin{pmatrix}1&0&0&\alpha_i\\0&1&0&0\\0&0&1&0\\0&0&0&1\end{pmatrix}.$$

5. Rotation is performed around the axis $X_{i-1}^{(3)}$ for the angle of α_i until the axis $Z_{i-1}^{(5)}$ is located in the plane passing through the line AB and axis Z_i. In this case, the axis $Y_{i-1}^{(5)}$ becomes perpendicular to the axis Z_i. This conversion is described by the rotation matrix

$$R\left(X_{i-1}^{(4)},\alpha_i\right)=\begin{pmatrix}1&0&0&0\\0&C\alpha_i&-S\alpha_i&0\\0&S\alpha_i&C\alpha_i&0\\0&0&0&1\end{pmatrix}.$$

6. The axis $Z_{i-1}^{(5)}$ is rotated relative to the axis $Y_{i-1}^{(5)}$ for the angle of $\psi_i=\frac{\pi}{2}-\lambda_i$ so that the directions of the axes $Z_{i-1}^{(6)}$ and Z_i coincide. The said action corresponds to the rotation matrix

$$R\left(Y_{i-1}^{(5)},\psi_i\right)=\begin{pmatrix}C\psi_i&0&S\psi_i&0\\0&1&0&0\\-S\psi_i&0&C\psi_i&0\\0&0&0&1\end{pmatrix}.$$

7. A transfer along the axis $Z_{i-1}^{(6)}$ is performed for the distance of b_i to match the center of the intermediate system $BX_{i-1}^{(6)}Y_{i-1}^{(6)}Z_{i-1}^{(6)}$ with the point O_i. This movement is described by the transfer matrix

$$T\left(Z_{i-1}^{(6)},b_i\right)=\begin{pmatrix}1&0&0&0\\0&1&0&0\\0&0&1&b_i\\0&0&0&1\end{pmatrix}.$$

8. Rotation is performed around the axis $Z_{i-1}^{(7)}$ for the angle of β_i to match the directions of the axes $X_{i-1}^{(7)}$ and X_i. This rotation corresponds to the rotation matrix

$$R\left(Z_{i-1}^{(7)},\beta_i\right)=\begin{pmatrix}C\beta_i&-S\beta_i&0&0\\S\beta_i&C\beta_i&0&0\\0&0&1&0\\0&0&0&1\end{pmatrix}.$$

After these eight movements, the coordinate systems $O_{i-1}\,X_{i-1}\,Y_{i-1}\,Z_{i-1,}$ and $O_i\,X_i\,Y_i\,Z_i$ fully coincide. Analytical representation of the composition of the given movements' sequence allows determining the required transformation matrix $\mathbf{A}_i^{i-1}$. Matrix $\mathbf{A}_i^{i-1}$ is found by sequential multiplying of the obtained by every movement homogeneous transformation matrices, i.e.,

$$\begin{aligned}\mathbf{A}_i^{i-1}(\delta_t,\theta_t,\gamma_t,\alpha_t,\alpha_t,\psi_t,\beta_t,\beta_t) = \boldsymbol{T}(Z_{t-1},\delta_t)\boldsymbol{R}\left(Z_{i-1}^{(1)},\theta_i\right)\boldsymbol{P}\left(X_{i-1}^{(2)},\gamma_t\right)\\ \boldsymbol{T}\left(X_{i-1}^{(3)},a_i\right)\boldsymbol{P}\left(X_{i-1}^{(4)},\alpha_v\right)\boldsymbol{P}\left(Y_{i-1}^{(5)},\psi_i\right)\\ \boldsymbol{T}\left(Z^{(6)},\beta_l\right)\boldsymbol{P}\left(Z_{i-1}^{(7)},\beta_l\right)\end{aligned} \tag{4.1}$$

Matrix $\mathbf{A}_i^{i-1}$ generally has the form of

$$\mathbf{A}_i^{i-1} = \begin{pmatrix} a_{11} & a_{12} & a_{13} & a_{14} \\ a_{21} & a_{22} & a_{23} & a_{24} \\ a_{31} & a_{32} & a_{33} & a_{34} \\ 0 & 0 & 0 & 1 \end{pmatrix} \tag{4.2}$$

Elements of the matrix $\mathbf{A}_i^{i-1}$ (4.2) are recorded in the form of the following expressions

$$\begin{aligned}
a_{11} &= C\beta C\theta C\gamma C\psi - C\beta S\theta S\alpha + C\beta S\psi C\theta S\gamma C\alpha - S\beta S\theta C\alpha \\
&\quad + S\beta C\theta S\gamma S\alpha;\\
a_{12} &= -S\beta C\theta C\gamma C\psi + S\beta S\theta S\alpha - S\beta S\psi C\theta S\gamma C\alpha - C\beta S\theta C\alpha \\
&\quad + C\beta C\theta S\gamma S\alpha;\\
a_{13} &= C\theta C\gamma S\psi + C\psi S\theta S\alpha + C\psi C\theta S\gamma C\alpha;\\
a_{14} &= b\cdot C\theta C\gamma S\psi + b\cdot C\psi S\theta S\alpha + b\cdot C\psi C\theta S\gamma C\alpha + a\cdot C\theta C\gamma;\\
a_{21} &= C\beta S\theta C\gamma C\psi + C\beta S\psi C\theta S\alpha - C\beta S\psi S\theta S\gamma C\alpha + S\beta C\theta C\alpha \\
&\quad + S\beta S\theta S\gamma S\alpha;\\
a_{22} &= -S\beta S\theta C\gamma C\psi - S\beta S\psi C\theta S\alpha + S\beta S\psi S\theta S\gamma C\alpha + C\beta C\theta C\alpha \\
&\quad + C\beta S\theta S\gamma S\alpha;\\
a_{23} &= S\theta C\gamma S\psi - C\psi C\theta S\alpha + C\psi S\theta S\gamma C\alpha;\\
a_{24} &= b\cdot S\theta C\gamma S\psi - b\cdot C\psi C\theta S\alpha + b\cdot C\psi S\theta S\gamma C\alpha + a\cdot S\theta C\gamma;\\
a_{31} &= -C\beta S\gamma C\psi + C\beta C\gamma C\alpha S\psi + C\gamma S\alpha S\beta;\\
a_{32} &= S\beta S\gamma C\psi + S\beta C\gamma C\alpha S\psi + C\gamma S\alpha C\beta;\\
a_{33} &= -S\gamma S\psi + C\gamma C\alpha C\psi;\\
a_{34} &= -b\cdot S\gamma S\psi + b\cdot C\gamma C\alpha C\psi - a\cdot S\gamma + d;\\
a_{41} &= a_{42} = a_{43} = 0;\\
a_{44} &= 1.
\end{aligned} \tag{4.3}$$

Comparison of homogeneous transformation matrices by four and six parameters and the obtained matrix $\mathbf{A}_i^{i-1}$ shows that the matrix of the D-H representation of four parameters is obtained from the matrix $\mathbf{A}_i^{i-1}$ with $\gamma_\iota = \psi_\iota = \beta_\iota = \beta_\iota = 0$, and matrix of six parameters—with $\gamma_\iota = \psi_\iota = 0$. But in the matrices, parameters $\gamma_\iota = \psi_\iota = \beta_\iota = \beta_\iota = 0$ or parameters $\gamma_\iota = \psi_\iota = 0$ are used in describing the special cases of the interconnection of two bodies. From this follows that the transformation matrix $\mathbf{A}_i^{i-1}$ with eight transformation parameters representing a more general description of the analytical description is acceptable for modeling mechanisms and manipulators with any kinematic pairs.

4.5 Examples of Redundant Parameters' Method Application

Let us consider the example of the mechanism shown in Fig. 4.3, which can be described by eight transformation parameters. Here, the fixed coordinate system is related to the support 0. Axis Z_0 is directed along the CEKP of the links 0–1. The point A is chosen at the intersection of the axes of 3 rotational pairs 1–2, 2–3, and 4–5. AB is the connecting straight line directed along the link 3 and 4. The point B is at the intersection of three rotational pairs 4–5, 5–6, and 6–01.

Axis Z_i is directed along the CEKP of the links 6–01. When transforming the i-coordinate system into i–1, transformation parameters of the matrix $\mathbf{A}_1^0$ will have values of the specific geometric and kinematic parameters of the links and KP, i.e., $d_i = s_{0-1}$, $\theta_i = \varphi_{1-2}$, $\gamma_i = \varphi_{2-3}$, $a_i = s_{3-4}$, $\alpha_i = \varphi_{4-5}$, $\psi_i = \varphi_{5-6}$, $b_i = O_1B$, $\beta_i = \varphi_{6-01}$. (Subindex

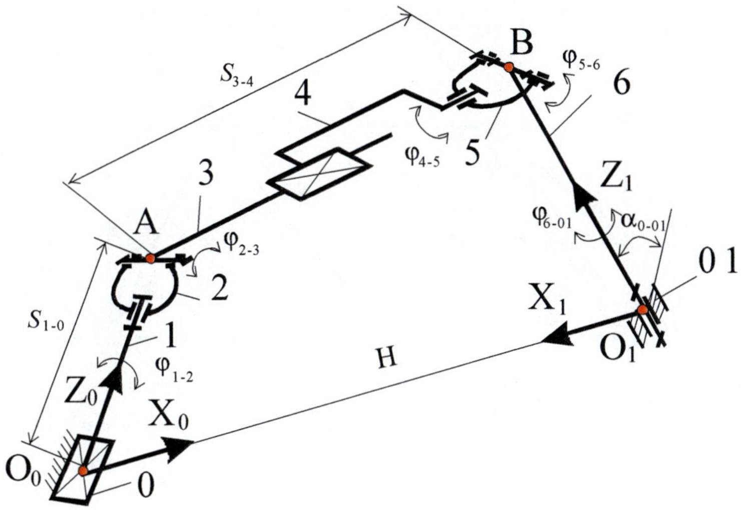

Fig. 4.3 Example of selecting the coordinate system and connecting straight line

of the parameters of the links and KP denotes links forming the KP.) As the kinematic chain is closed, another transformation of the system 0 into the system 1 through the line O_0O_1 is required. To simplify the problem, position of the supports 0 and 01 is selected so that O_0O_1 are a common perpendicular to the axis Z_0 and Z_1. Therefore, in this transformation, most of the elements of the transformation matrix $\mathbf{A}_0^1$ turn into zero. The zero parameters are $a_i = H$ and the twist angle of the axes $\alpha_i = \varphi_{01-0}$. The expression for the elements of the transformation matrix (4.3) inserts specific values of links and KP parameters.

Further, there is a solution for the matrix equation

$$\mathbf{A}_1^0 \cdot \mathbf{A}_0^1 = \boldsymbol{E}, \tag{4.4}$$

where $\boldsymbol{E}$ is a unit matrix 4×4.

Since the mechanism shown in Fig. 4.3 has the DOF equal to one, i.e., it is described by one generalized coordinate, the algebraic system of equations obtained from the matrix Eq. (4.3) determines dependencies for the kinematic parameters of the generalized coordinate. When applying the method [42], the matrix expression would be represented by an expression of 10 matrices. Another example of a spatial 4-link mechanism illustrated in Fig. 4.4 makes it possible to demonstrate that the use of the proposed mathematical model helps to take into account, for example, the arrangement of the slot plane in the two-DOF spherical pair. In this mechanism, specifically the slot planes in the two-DOF spherical pairs are located in one plane (the plane of the drawing). This allows comparing the results of the solution obtained for the mechanism in question with known results for a similar flat lever mechanism with

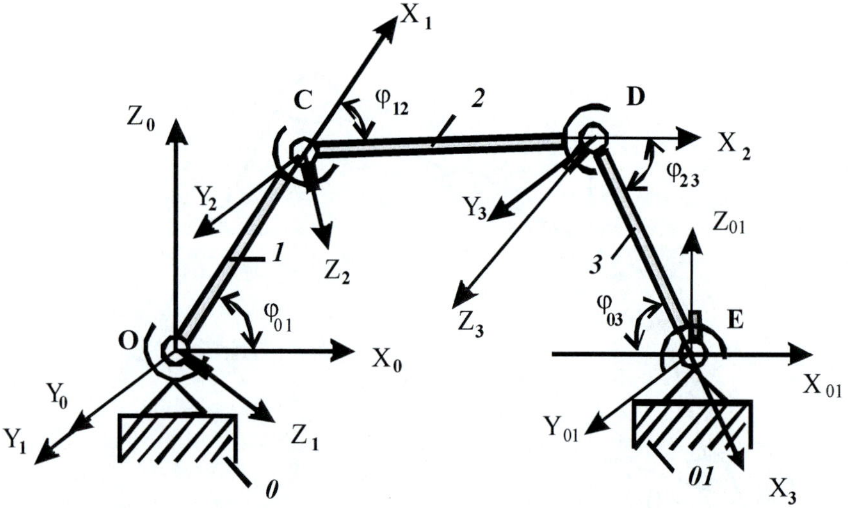

Fig. 4.4 Spatial mechanism

rotational KP. This example also shows that the present method allows for relating each link to the coordinate system, given the characteristics of the KP when it is required by the mechanism's design. In accordance with the above recommendations, the coordinate systems related to the links are chosen. In the fixed coordinate systems, the axes Y_0 and Y_{01} are directed perpendicular to the planes of the slots, axis Z_{01}—along the pin. Since there is freedom in choosing axes related to the supports, then it is provided that the axes X_0, Z_0 and axes X_{01}, Z_{01} are mutually parallel. To compile a matrix equation of the closed kinematic chain, it is provided to the transformation of the fixed system 0 into the system 01 related to the support. Therefore, coordinate axis related to the supports is specifically chosen so that transformation of the coordinate system $O_0\, X_0\, Y_0\, Z_0$ into $O_{01}\, X_{01}\, Y_{01}\, Z_{01}$ is performed with a minimum number of the transformation parameters.

Coordinate axes Z_1, Z_2, and Z_3 in accordance with the recommendations are directed along the pins away from the KP; axes X_1, X_2 and X_3 are directed along the link, and axes Y_1, Y_2, and Y_3—perpendicular to the plane of the slot (drawing plane).

The following notation is identified: $OC = L_1$, $CD = L_2$, $DE = L_3$ are lengths of moving links; $OE = L_{01}$ is the shortest distance between the supports; $\varphi_{01}, \varphi_{12}, \varphi_{23}$, and φ_{03} are the angles between the links, variable parameters of the KP defined as the angle between the positive directions of the axis X.

After selecting the axes for each KP, specific transformation parameters are set, which are summarized in Table 4.1. Having substituted the transformation matrix elements with the corresponding data from Table 4.1, the transformation matrix $\mathbf{A}_i^{i-1}$ is formed. Then the matrix equation is made for the closed loop in the form of

$$\mathbf{A}_1^0 \cdot \mathbf{A}_2^1 \cdot \mathbf{A}_3^2 \cdot \mathbf{A}_{01}^3 \cdot \mathbf{A}_0^{01} = \boldsymbol{E}. \tag{4.5}$$

It should be noted that the use of the known matrix methods did not allow describing the features of the KP of such a mechanism, as well as producing a matrix expression with five multipliers. Typically, a problem solution is conveniently divided into two stages. At the first stage, the necessary transformation matrix is determined,

Table 4.1 Transformation matrix elements

Coordinate systems	Notation of the transformation matrix	Transformation parameters							
		d_i	θ_i	γ_i	a_i	α_i	ψ_i	b_i	β_i
0	1	2	3	4	5	6	7	8	9
0–1	$\mathbf{A}_1^0$	0	φ_{01}	0	0	0	$\pi - \varphi_{01}$	0	0
0	1	2	3	4	5	6	7	8	9
1–2	$\mathbf{A}_2^1$	0	0	0	L_1	0	φ_{12}	0	0
2–3	$\mathbf{A}_3^2$	0	0	0	L_2	0	$-\varphi_{23}$	0	0
3–01	$\mathbf{A}_{01}^3$	0	0	0	L_3	0	$-\varphi_{30}$	0	0
01–0	$\mathbf{A}_0^{01}$	0	0	0	L_{01}	0	0	0	0

and depending on the design, the analytic description is given, i.e., a mathematical model of the mechanism or manipulator is created. Here, the adequacy of the mathematical model and its complexity depends on the chosen method of analytical description. The proposed method allows constructing an adequate model for any mechanism and manipulator.

At the second stage, the search for solutions of transcendental equation systems depends on the matrix expressions. The proposed method provides more simple final matrix expressions, and thereby, facilitates the solution of the equations, which determine the kinematic parameters of mechanisms and manipulators.

Part III
Kinematics and Dynamics of One-Loop and Multiloop Parallel Manipulators

Chapter 5
Kinematics of One-Loop Parallel Manipulators

Abstract This chapter provides kinematic analysis of OPM that is based on CFG kinematics analysis results. The direct and inverse kinematics problems of OPM are solved. For further dynamic analysis, kinematic parameters are obtained for OPM links in the coordinate systems related to the links. Homogeneous matrix method is obtained for determining velocities and accelerations of the OPM links.

Keywords Core Functional Groups · Velocities · Accelerations · Direct and inverse problems of OPM

5.1 Homogeneous Transformation Matrices for Core Functional Groups

The mechanism, where the output link is related to the base by several kinematic chains, each of which contains one or more drives or imposes a number of connections on the output link, is called a mechanism with parallel structure or parallel manipulator.

As follows from its definitions, the mechanisms of the parallel manipulators are one- and multiloop multi-degree-of-freedom spatial mechanisms. These mechanisms are described in the studies of foreign and domestic authors [1, 3, 43, 45]. This chapter examines the parallel one-loop manipulators proposed by the author for the first time in 1982 [13, 14], and presents the modified theory of one-loop parallel manipulators described previously in the study [43]. One-loop parallel manipulators are multi-degree-of-freedom spatial mechanisms with closed kinematic chains. Kinematic analysis of the manipulators mainly solves the following tasks:

- determine linear and angular displacements of the links and various path of the points;
- determine the velocity of individual points and the angular velocity of the links;
- determine the acceleration of the individual points and the angular acceleration of the links.

K. S. Sholanov, *Parallel Manipulators of Robots*, Mechanisms and Machine Science 92,
https://doi.org/10.1007/978-3-030-56073-7_5

A feature of the kinematics of this manipulators is that the kinematics of the manipulators distinguishes between direct and inverse kinematics problem. The kinematics of the manipulators also solves the problem of determining the kinematic positioning errors, constructing maintenance zones, and etc. The basic mathematical apparatus for the study are those utilizing the matrix methods using homogeneous transformation matrices.

For one-loop parallel manipulators consisting of complete or incomplete CFG, it appears to be convenient to use the method of Denavit-Hartenberg [29] (D-H representation).It is known that the scope of application of the D-H representation is limited only by the manipulator mechanisms with lower pairs and with a certain (parallel, perpendicular) orientation of the kinematic pairs axes (as in CFG). Below, it is assumed that the kinematic chains of one-loop parallel manipulators are formed by interconnection of the CFG or incomplete CFG and the connecting link. While the kinematic model of the manipulator in the form of a homogeneous transformation matrix is obtained using transformation matrices of CFG and CL.

Below, using D-H representation, homogeneous transformation matrix of CFG consisting of three parts are obtained, i.e. CFG matrix. It is obvious that CFG matrix can be obtained as a product of matrices of the kinematic pairs.

It is, therefore, necessary to obtain matrices of the kinematic pairs for each CFG.

In this case, five CFG were considered; they had no hollow areas in the maintenance zone, the maintenance zone dimensions were determined by the generalized coordinates. Each of CFG links (Fig. 5.1) is related to a rectangular coordinate system, while the direction and position of the coordinate axes are selected according to the D-H representation. It should be noted that in the CFG, axes of the rotational kinematic pairs and directing translational kinematic pairs are mutually perpendicular, parallel or coincide, i.e. Represent the simplest orientation. In this case, the use of D-H presentation is obvious. Thus, for each of the kinematic pairs, corresponding parameters of the D-H transformation are determined: θ_ι, σ_ι, α_ι,α_ι. The values of the transformation parameters are summarized in Table 5.1 named table of the kinematic pairs. Designations for generalized coordinates are accepted according to Fig. 5.1. Let us substitute the appropriate parameters into Table 5.1.

Extended matrix for *i*-th kinematic pair has the following form

$$\mathbf{A}_i = \begin{pmatrix} C\theta_i & -C\alpha_i S\theta_i & S\alpha_i S\theta_i & a_i C\theta_i \\ S\theta_i & C\alpha_i C\theta_i & -S\alpha_i C\theta_i & a_i S\theta_i \\ 0 & S\alpha_i & C\alpha_i & s_i \\ 0 & 0 & 0 & 1 \end{pmatrix}, \tag{5.1}$$

where abbreviations are introduced for the functions: *Cos* – *C, Sin* – *S.*

Substituting the values of the transformation parameters from Table 5.1 in the expression (5.1), the matrix of kinematic pairs of each CFG is obtained in the following form:

CFG-1

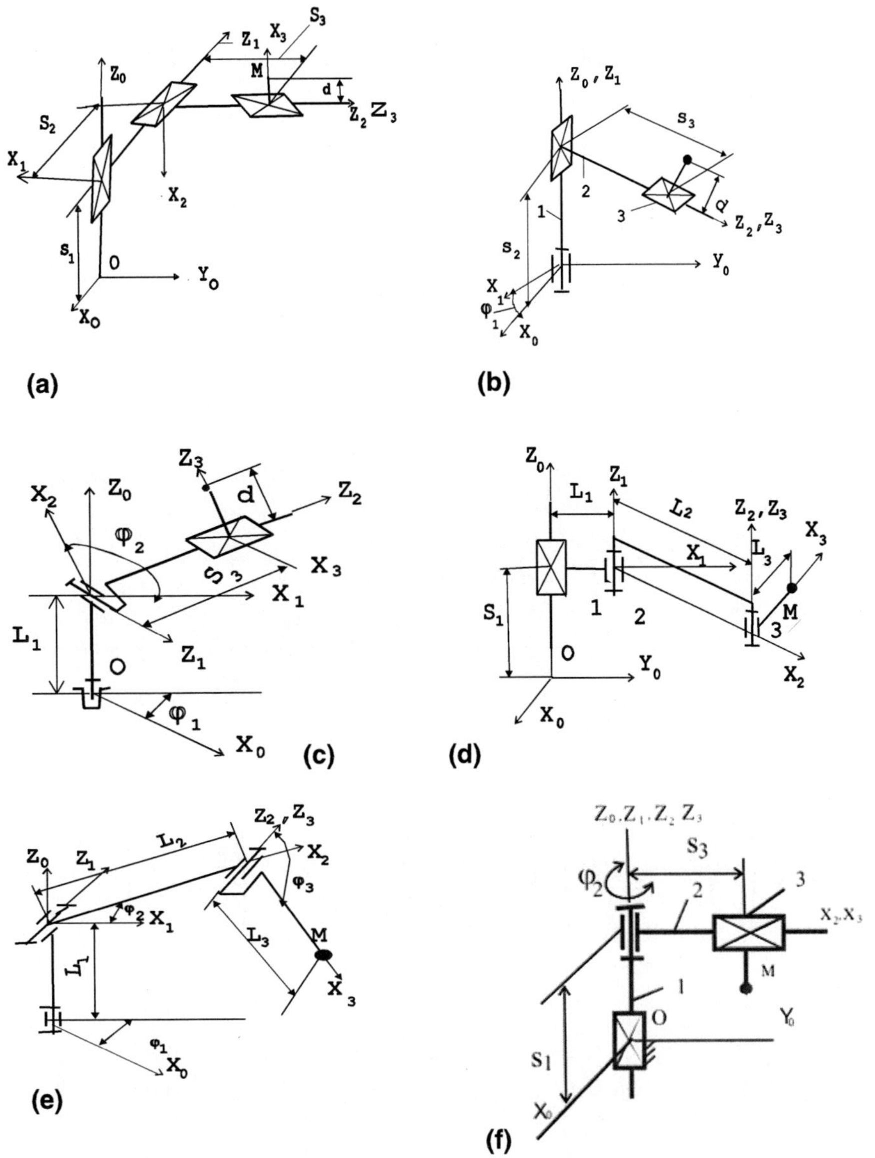

Fig. 5.1 Selection of CFG coordinate axes

$$\mathbf{A}_1 = \begin{pmatrix} 0 & 0 & 1 & 0 \\ 1 & 0 & 0 & 0 \\ 0 & 1 & 0 & s_1 \\ 0 & 0 & 0 & 1 \end{pmatrix} \mathbf{A}_2 = \begin{pmatrix} 0 & 0 & 1 & 0 \\ 1 & 0 & 0 & 0 \\ 0 & 1 & 0 & s_2 \\ 0 & 0 & 0 & 1 \end{pmatrix} \mathbf{A}_3 = \begin{pmatrix} -1 & 0 & 0 & 0 \\ 0 & 1 & 0 & 0 \\ 0 & 0 & 1 & s_3 \\ 0 & 0 & 0 & 1 \end{pmatrix}; \tag{5.2}$$

Table 5.1 Values of the transformation parameters for CFG

No. CFG	Links forming KP	Name of KP	Transformation parameters			
			d_i	θ_i	a_i	α_i
1	0–1 1–2 2–3	Transl.Transl. Transl.	s_1 s_2 s_3	$\pi/2$ $\pi/2$ π	0 0 0	$\pi/2$ $\pi/2$ 0
2	0–1 1–2 2–3	Rotat. Transl.Const.	0 s_2 s_3	φ_1 π 0	0 0 0	0 $\pi/2$ 0
3	0–1 1–2 2–3	Rotat. Rotat. Transl.	L_1 0 s_3	φ_1 φ_2 0	0 0 0	$\pi/2$ $\pi/2$ $\pi/2$
4	0–1 1–2 2–3	Transl. Rotat. Rotat.	s_1 0 0	$\pi/2$ $-\varphi_2$ φ 3	L_1 L_2 0	0 0 0
5	0–1 1–2 2–3	Rotat. Rotat. Rotat.	L_1 0 0	φ_1 $-\varphi_2$ φ_3	0 L_2 0	$-\pi/2$ 0 0
6	0–1 1–2 2–3	Transl.Rotat. Transl	s_1 0 0	0 φ_2 0	0 0 s_3	0 0 0

CFG-2

$$\mathbf{A}_1 = \begin{pmatrix} C\phi_1 & -S\phi_1 & 0 & 0 \\ S\phi_1 & C\phi_1 & 0 & 0 \\ 0 & 0 & 1 & 0 \\ 0 & 0 & 0 & 1 \end{pmatrix}, \mathbf{A}_2 = \begin{pmatrix} -1 & 0 & 0 & 0 \\ 0 & 0 & 1 & 0 \\ 0 & 1 & 0 & s_2 \\ 0 & 0 & 0 & 1 \end{pmatrix}, \mathbf{A}_3 = \begin{pmatrix} 1 & 0 & 0 & 0 \\ 0 & 1 & 0 & 0 \\ 0 & 0 & 1 & s_3 \\ 0 & 0 & 0 & 1 \end{pmatrix}; \tag{5.3}$$

CFG-3

$$\mathbf{A}_1 = \begin{pmatrix} C\phi_1 & 0 & S\phi_1 & 0 \\ S\phi_1 & 0 & -C\phi_1 & 0 \\ 0 & 1 & 0 & L_1 \\ 0 & 0 & 0 & 1 \end{pmatrix}, \mathbf{A}_2 = \begin{pmatrix} C\phi_2 & 0 & S\phi_2 & 0 \\ S\phi_2 & 0 & -C\phi_2 & 0 \\ 0 & 1 & 0 & 0 \\ 0 & 0 & 0 & 1 \end{pmatrix}, \mathbf{A}_3 = \begin{pmatrix} 0 & 0 & 1 & 0 \\ 0 & 0 & -1 & 0 \\ 0 & 1 & 0 & s_3 \\ 0 & 0 & 0 & 1 \end{pmatrix}; \tag{5.4}$$

CFG-4

$$\mathbf{A}_1 = \begin{pmatrix} 0 & -1 & 0 & 0 \\ 1 & 0 & 0 & L_1 \\ 0 & 0 & 1 & s_1 \\ 0 & 0 & 0 & 1 \end{pmatrix}, \mathbf{A}_2 = \begin{pmatrix} C\phi_2 & S\phi_2 & 0 & L_2C\phi_2 \\ -S\phi_2 & C\phi_2 & 0 & -L_2S\phi_2 \\ 0 & 0 & 1 & 0 \\ 0 & 0 & 0 & 1 \end{pmatrix}, \mathbf{A}_3 = \begin{pmatrix} C\phi_3 & -S\phi_3 & 0 & 0 \\ S\phi_3 & C\phi_3 & 0 & 0 \\ 0 & 0 & 1 & 0 \\ 0 & 0 & 0 & 1 \end{pmatrix}; \tag{5.5}$$

CFG-5

$$\mathbf{A}_1 = \begin{pmatrix} C\phi_1 & 0 & -S\phi_1 & 0 \\ S\phi_1 & 0 & C\phi_1 & 0 \\ 0 & -1 & 0 & L_1 \\ 0 & 0 & 0 & 1 \end{pmatrix}, \mathbf{A}_2 = \begin{pmatrix} C\phi_2 & 0 & S\phi_2 & L_2C\phi_2 \\ -S\phi_2 & 0 & C\phi_2 & -L_2S\phi_2 \\ 0 & 0 & 1 & 0 \\ 0 & 0 & 0 & 1 \end{pmatrix},$$

$$\mathbf{A}_3 = \begin{pmatrix} C\phi_3 & -S\phi_3 & 0 & 0 \\ S\phi_3 & C\phi_3 & 0 & 0 \\ 0 & 0 & 1 & 0 \\ 0 & 0 & 0 & 1 \end{pmatrix}. \tag{5.6}$$

CFG-6

$$\mathbf{A}_1 = \begin{pmatrix} 1 & 0 & 0 & 0 \\ 0 & 1 & 0 & 0 \\ 0 & 0 & 1 & s_2 \\ 0 & 0 & 0 & 1 \end{pmatrix}, \mathbf{A}_2 = \begin{pmatrix} C\phi_2 & -S\phi_2 & 0 & 0 \\ S\phi_2 & C\phi_2 & 0 & 0 \\ 0 & 0 & 1 & 0 \\ 0 & 0 & 0 & 1 \end{pmatrix}, \mathbf{A}_3 = \begin{pmatrix} 1 & 0 & 0 & s_3 \\ 0 & 1 & 0 & 0 \\ 0 & 0 & 1 & 0 \\ 0 & 0 & 0 & 1 \end{pmatrix}. \tag{5.7}$$

The matrix of the homogeneous transformation of the CFG – $\boldsymbol{T}_i$ ($i = 1 \div 6$) is obtained by multiplying the matrices of the kinematic pairs for each individual CFG, i.e.

$$\boldsymbol{T}_i = \boldsymbol{A}_1\boldsymbol{A}_2\boldsymbol{A}_3. \tag{5.8}$$

Here $\boldsymbol{A}_1$, $\boldsymbol{A}_2$, $\boldsymbol{A}_3$ are the matrices of the kinematic pairs of the i-th CFG formed respectively by the links 0–1, 1–2, 2–3, expressions for which are taken from the formulas (5.2)–(5.7).

5.2 Direct Position Problem for the CFG

Obtained with the help of (5.8) of the matrices of the CFG are the basis for solving the direct kinematics problem for CFG. Relative to the reference coordinate system, position of the characteristic point M of the link that is last from the base, for each of the CFG is determined generally by the vector equation

$$\mathbf{r}_{\mathrm{M}} = \mathbf{T}_i\mathbf{r}_M^3, \tag{5.9}$$

where $\boldsymbol{r}_M = (x_M, y_M, z_M, 1)^T$ is the augmented vector of the homogeneous coordinates of the point M relative to the reference coordinate system; $\boldsymbol{r}_M^3 = \left(x_M^3, y_M^3, z_M^3, 1\right)^T$ is the vector composed of homogeneous coordinates of the point M relative to the coordinate system O_3, X_3, Y_3, Z_3 related to the link 3.

The coordinates of the characteristic point M of the last link of the CFG $\left(x_M^3, y_M^3, z_M^3\right)$ are taken from the design considerations. In our case, they follow from the kinematics CFG (Fig. 5.1).Then the vector $\boldsymbol{r}_M^3$ for the CFG-i ($i = 1 \div 6$) has the following values: (1) $\boldsymbol{r}_M^3 = (d\ 0\ 0\ 1)^T$; (2) $\boldsymbol{r}_M^3 = (0\ d\ 0\ 1)^T$; (3) $\boldsymbol{r}_M^3 = (0\ 0\ d\ 1)^T$; (4) $\boldsymbol{r}_M^3 = (L_3\ 0\ 0\ 1)^T$; (5) $\boldsymbol{r}_M^3 = (L_3\ 0\ 0\ 1)^T$; (6) $\boldsymbol{r}_M^3 = (s\ 0\ -d\ 1)$.

Having substituted the expressions for the CFG homogeneous transformation matrices (5.2–5.7) into the formula (5.9) and using appropriate to CFG values of $\boldsymbol{r}_M^3$, coordinates of the point M of the output link are obtained in the reference coordinate system

CFG-1:

$$x_M = s_2;\ y_M = s_3;\ z_M = s_1 - d. \tag{5.10}$$

CFG-2:

$$x_M = -s_3 S\varphi_1;\ y_M = s_3 C\varphi_1;\ z_M = d + s_2. \tag{5.11}$$

CFG-3:

$$x_M = s_3 S\varphi_2 C\varphi_1 - d S\varphi_1;\ y_M = s_3 S\varphi_2 S\varphi_1 + d C\varphi_1;\ z_M = L_1 - s_3 C\varphi_2. \tag{5.12}$$

CFG-4:

$$x_M = L_3 S(\varphi_2 - \varphi_3) + L_2 S\varphi_2;\ y_M = L_3 C(\nu_2 - \nu_3) + L_2 C\nu_2 + L_1;\ z_M = S \tag{5.13}$$

CFG-5:

$$\begin{aligned} x_M = {} & L_3 C\varphi_1 C(\varphi_2 - \varphi_3) + L_2 C\varphi_2 C\varphi_1;\ y_M = L_3 S\varphi_1 C(\varphi_2 - \varphi_3) \\ & + L_2 S\varphi_1 C\varphi_2;\ z_M = L_3 S(\varphi_2 - \varphi_3) + L_2 S\varphi_2 + L_1. \end{aligned} \tag{5.14}$$

CFG-6:

$$x_M = 2 s_3 C\varphi_2;\ y_M = 2 s_3 S\varphi_2;\ z_M = s_1 - d. \tag{5.15}$$

The Eqs. (5.10–5.15) establish the dependence between the coordinates of the point of the CFG output link and generalized coordinates, i.e. are the function of the CFG position. The resulting formulas allow determining the maintenance zone for the manipulator composed of the CFG with the set limits of changing the generalized coordinates and various design parameters. Furthermore, the obtained dependences are the initial ones for solving the inverse problem for CFG.

5.3 Inverse Position Problem for the CFG

It is known that the inverse kinematics position problem defines generalized coordinates of the manipulator with predetermined positions of the output link. In multi-degree-of-freedom mechanisms of robots having the open kinematic chain, the inverse problem of the positions and cannot be solved unambiguously and explicitly. Otherwise, mechanisms of the one-loop parallel manipulators (OPM) provide an opportunity to explicitly solve direct and reverse position problems. Since OPMs consist of CFG, the solution of the inverse position problem is initially discussed here in relation to the base manipulators formed of CFG. Then the results obtained for the CPG can be used to solve the inverse position problem for the OPM as a whole.

Let us consider sequentially the solution of the inverse kinematics problem for the base manipulators formed of the previously discussed six CFGs. In the manipulator formed by the CFG-1 (Fig. 5.1a), the predetermined position of the characteristic point M in the reference system is determined by the coordinates x_M, y_M, z_M. The Eq. (5.10) gives the following generalized coordinates

$$s_1 = z_M + d;\ s_2 = x_M;\ s_3 = y_M. \tag{5.16}$$

Generalized coordinates of the CFG-2 according to the predetermined coordinates of the point M are identified after the simple transformation from the Eq. (5.11)

$$\phi_1 = -\mathrm{arctg}\frac{x_M}{y_M}\ ;\ s_2 = z_M - d\ ;\ s_3 = \sqrt{x_M^2 + y_M^2}. \tag{5.17}$$

It is assumed that the range of variation of the angle φ_1 corresponds to $-\pi/2 < \varphi_1 < -\pi/2$. From the physical sense s_3 and the choice of the coordinate system, it follows that s_3 is a positive variable.

After transformations of the equality (5.12), which are omitted because of their cumbersome, generalized coordinates are obtained for CFG-3

$$\phi_1 = \arcsin\frac{-x_M d + y_M\sqrt{x_M^2 + y_M^2 - d^2}}{x_M^2 + y_M^2};$$

$$\phi_2 = \arccos\frac{L_1 - z_M}{\sqrt{x_M^2 + y_M^2 + (z_M - L_1)^2 + d^2}};$$

$$s_3 = \sqrt{x_M^2 + y_M^2 + (z_M - L_1)^2 + d^2} \tag{5.18}$$

Dependence for generalized coordinates of the CFG-4 is obtained from the system of Eqs. (5.13). While also, as it is done above, the intermediate steps are omitted.

$$s_1 = z_M; \phi_2 = \arcsin\frac{(y_M - L_1)^2 - L_3^2 + x_M^2 + L_2^2}{2L_2\sqrt{x_M^2 + (y_M - L_1)^2}} - \mathrm{arctg}\frac{y_M - L_1}{x_M};$$

$$\phi_3 = \arcsin\frac{(y_M - L_1)^2 - L_3^2 + x_M^2 + L_2^2}{2L_2\sqrt{x_M^2 + (y_M - L_1)^2}} - \mathrm{arctg}\frac{y_M - L_1}{x_M}$$

$$+ \arcsin\left(\frac{\frac{x_M}{L_3} - \frac{L_2}{L_3}\frac{x_M\left((y_M - L_1)^2 - L_3^2 + x_M^2 + L_2^2\right)}{2L_2\left(x_M^2 + (y_M - L_1)^2\right)}}{-\frac{y_M - L_1}{\sqrt{4L_2^2\left(x_M^2 + (y_M - L_1)^2\right) - \left((y_M - L_1)^2 - L_3^2 + x_M^2 + L_2^2\right)}}}\right). \tag{5.19}$$

The inverse problem solution for CFG-5 follows the solutions of the systems of Eqs. (5.14)

$$\begin{aligned}
&\phi_1 = \mathrm{arctg}\frac{y_M}{x_M}; \phi_2 = \arccos\frac{-bc + a\sqrt{a^2 + b^2 - c^2}}{a^2 + b^2},\\
&a = x_M^2 + y_M^2 + (z_M - L_1)^2 + L_2^2 - L_3^2,\\
&b = \sqrt{4L_2^2L_3^2 - x_M^2 + y_M^2 + (z_M - L_1)^2 - L_2^2 - L_3^2}, c = (z_M - L_1)2L_2;\\
&s_3 = \arccos\frac{x_M^2 + y_M^2 - L_3^2 - (z_M - L_1)^2 - L_2^2}{2L_2L_3}.
\end{aligned} \tag{5.20}$$

For the manipulator of the CFG-6, inverse problem solution is obtained from the equalities (5.15) as follows:

$$\begin{aligned}
s_1 &= z_M + d,\\
s_3 &= {}^1/_2\sqrt{x_M^2 + y_M^2},\\
\phi_2 &= arc\mathrm{Cos}\left({}^{x_M}/_{2s_3}\right).
\end{aligned} \tag{5.21}$$

The obtained dependences for the CFG (5.18–5.21) are inverse solution for the kinematics problem of the position for the 3-DOF moving manipulators composed of CFG.

5.4 Solution of the Direct Position Problem for the OPM

The solution of the direct position problem for the OPM is explained on the example of the manipulator (Fig. 5.3) formed by two CFG-3 according to the method described in paragraph 5.2. In the manipulator in Fig. 5.2, the kinematic pairs 0–1, 1–2, 2–

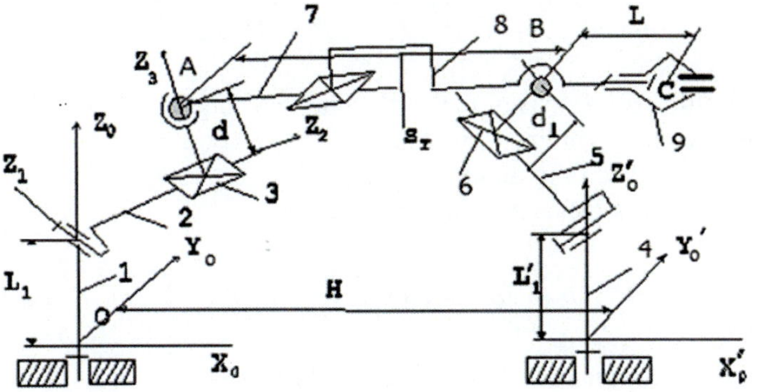

Fig. 5.2 Kinematic diagram of the OPM

3, 4–0, 4–5, 5–6, 8–9 are driving, i.e. movement of the links forming these pairs is performed using the drives. On the example of the selected manipulator, an algorithm is obtained for solving the direct kinematics problem. The support of one of the CFG (the first CFG) is related to the coordinate system $OX_0Y_0Z_0$, which is called reference coordinate system. The support of the second CFG comprised in the manipulator is related to the coordinate system $O'X'_0Y'_0Z'_0$ that is called auxiliary. Since there is a certain freedom in choosing the direction of the axis $O'X'_0$, the said axis is parallel to the axis OX_0. With this choice of the auxiliary coordinate system, the reference system may be combined with the auxiliary one by the only movement - transfer along the axis OX_o for the constant value *H*. Coordinates of the point *A* of the connecting link are determined as the coordinates of the characteristic point of the output link of the CFG-3 by formulas (5.12) and are equal to

$$X_A = s_3 S\varphi_2 C\varphi_1 - dS\varphi_1,\ Y_A = s_3 S\varphi_2 S\varphi_1 + dC\varphi_1,\ Z_A = L_1 - s_3 C\varphi_2, \quad (5.22)$$

where X_A, Y_A, Z_A are coordinates of the point *A* relative to the reference coordinate system.

The coordinates of the characteristic point *B* belonging to the CFG-3 are determined relative to the coordinate system $O'X'_0Y'_0Z'_0$ and by the formula (5.12). Then, with respect to the reference system, coordinates of the point *B* of the connecting link equal to

$$X_B = H + s_6 S\varphi_5 C\varphi_4 - d_1 S\varphi_5,\ Y_B = s_6 S\varphi_5 S\varphi_4 - d_1 C\varphi_4,$$
$$Z_B = L_1^1 - s_6 C\varphi_5, \quad (5.23)$$

where $\varphi_4, \varphi_5, s_6$ are generalized coordinates of the second CFG.

Let us determine the distance s_r between the points *A and B* of the connecting link. In the Fig. 5.2 it is seen that

$$s_r = \sqrt{\left((X_A - X_B)^2 + (Y_A - Y_B)^2 + (Z_A - Z_B)^2\right)}. \tag{5.24}$$

The orientation of the axis of the connecting link and the gripper relative to the axes of the reference coordinate system, with known coordinates of the two points of the line, is identified from the expression for direction cosines.

The angle between the connecting link and the axis OX is

$$\alpha = \arccos \frac{X_B - X_A}{s_r}. \tag{5.25}$$

The angle between the connecting link and the axis OY is

$$\beta = \arccos \frac{Y_B - Y_A}{s_r}. \tag{5.26}$$

The angle between the line AB and the axis OZ is

$$\gamma = \arccos \frac{z_B - z_A}{s_r}. \tag{5.27}$$

It should be noted that it is sufficient to determine only two angles, since one of the angles, with other angles known may be determined from the orthogonality relation.

$$\mathrm{Cos}\,\alpha^2 + \mathrm{Cos}\,\beta^2 + \mathrm{Cos}\,\gamma^2 = 1.$$

Coordinates of the gripper center *C* with the known coordinates of the points *A* and *B* are identified according to the theorem of the division of a segment in this ratio. The connecting link AC is divided by the point B in the ration of $\lambda = s_r \backslash L$.Then the coordinates of the point C of the gripper center are determined by the following expressions

$$X_c = \frac{L(X_B - X_A)}{s_r} + X_B;\ Y_c = \frac{L(Y_B - Y_A)}{s_r} + Y_B;$$
$$Z_c = \frac{L(Z_B - Z_A)}{s_r} + Z_B. \tag{5.30}$$

Here the values of the coordinates of the points *A* and *B*, and the center-to-center distance s_r, are found according to the formulas (5.22–5.24). Thus, the example of the one-loop manipulator c parallel design having the closed kinematic chain describes the method for determining the position of the gripper center $\boldsymbol{r_c} = (X_c, Y_c, Z_c, 1)^T$ and orientation of the manipulator connecting link. This method can be applied to any OPM composed of two complete and incomplete CFGs.

5.5 Velocities, Accelerations of the Points and Angular Velocities, Angular Accelerations of the Links

5.5.1 *Vector Method of Determining the Kinematic Parameters of the Manipulator Links*

Kinematic parameters defined in this section are used in the future in solving problems of dynamics. Therefore, the mathematical apparatus used to determine the velocities, accelerations and angular velocities of the points, angular accelerations of the manipulator links, is directly related to the subsequent dynamics analysis and synthesis methods. Let us consider: vector and matrix methods for determining the said kinematic parameters.

The vector method of studying kinematics uses the velocity addition theorem, solid bodies movement addition theorem, as well as the method of reducing the velocity vectors and angular velocities.

To derive the kinematic relations by the vector method, the coordinate system shown in Fig. 5.3 is chosen. Here the coordinate axis related to the *i-1, i, i + 1* manipulator links, are directed according to the accepted method of mathematical description of the related bodies by the D-H representation. Center of the *i*-th coordinate system is the point O_i, it performs a complex motion. To determine the velocity of O_i, let us form the vector equality

Fig. 5.3 Coordinate systems

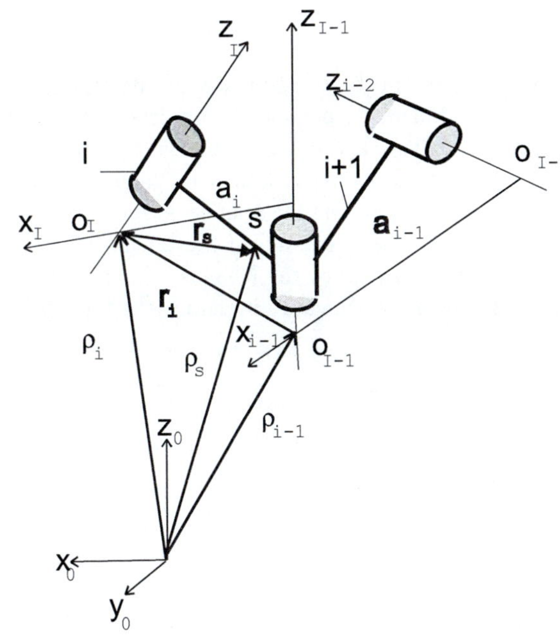

$$\bar{\rho}_i = \bar{\rho}_{i-1} + \bar{r}_i. \tag{5.31}$$

In it $\bar{\rho}_i$, $\bar{\rho}_{i-1}$ are radius vectors defining the position of respectively i and i-1 coordinate systems relative to the fixed coordinate system,; $\bar{r}_i$ is a radius vector describing the position of the coordinate systems relative to each other. Velocity of the point O_i is determined by the equality (5.31) differentiation taking into account the fact that $\bar{r}_i$-is a vector with alternating direction. Then

$$\bar{v}_i = \bar{v}_{i-1} + \bar{\omega}_{i-1} \times \bar{r}_i + \bar{v}_i^r. \tag{5.32}$$

The first two summands in (5.32) is the velocity of the translational motion of the point O_i, and the last summand is the relative velocity of the point O_i relative to the coordinate system $O_{i-1}X_{i-1}Y_{i-1}Z_{i-1}$. Angular velocity of the i-coordinate system $(\bar{\omega}_i)$ It is defined as the geometric sum of the angular velocity in the translational motion of the i-1 coordinate system relative to the fixed system ($\bar{\omega}_{i-1}$) and the angular velocity of own rotation of the i-coordinate system ($\bar{\omega}_i^r$).

$$\bar{\omega}_i = \bar{\omega}_{i-1} + \bar{\omega}_i^r. \tag{5.33}$$

The further differentiation of the Eqs. (5.32) (5.33) determine respectively them acceleration $\bar{a}_i$ of the point O_i and angular acceleration of the i-link $\bar{\varepsilon}_i$

$$\begin{aligned} \bar{a}_i &= \bar{a}_{i-1} + \bar{\varepsilon}_{i-1} \times \bar{r}_i + 2\bar{\omega}_{i-1} \times \bar{v}_i^r + \bar{\omega}_{i-1} \times (\bar{\omega}_{i-1} \times \bar{r}_i) + \bar{a}_i^r, \\ \bar{\varepsilon}_i &= \bar{\varepsilon}_{i-1} + \bar{\omega}_{i-1} \times \bar{\omega}_{i-1}^r + \bar{\varepsilon}_i^r. \end{aligned} \tag{5.34}$$

In the transition from the *i*-th to the *i + 1*-th coordinate system, *i* in the formula (5.32–5.34) should be substituted with the index *i + 1* instead of *i*. At the same time, *i − 1* kinematic parameters are determined by the results of calculations of the previous stage, and in the first stage of computing with *i = 1* in the formulas (5.32–5.34) initial values of kinematic parameters are used. Relative velocities $\bar{v}_i^r$, angular velocities $\bar{\omega}_i^r$ and the relative accelerations $\bar{a}_i^r$ and angular accelerations $\bar{\varepsilon}_i^r$ are expressed via the generalized coordinates of the manipulator depending on the kinematic pair formed by the *i- and i-1* links. The expressions for these values of different kinematic pairs are summarized in Table 5.2.

Table 5.2 Generalized coordinates

Kinematic parameters	Translational kin. pair	Rotational kin. pair
$\bar{\omega}_i^r$	0	$\bar{z}_{i-1}\dot{\theta}_i$
$\bar{v}_i^r$	$\bar{z}_{i-1}\dot{\theta}_i$	$\bar{\omega}_i^r \times \bar{r}_i$
$\bar{\varepsilon}_i^r$	0	$\bar{z}_{i-1}\ddot{\theta}_i$
$\bar{a}_i^r$	$\bar{z}_{i-1}\ddot{\theta}_i$	$\bar{\varepsilon}_i \times \bar{r}_i + \bar{\omega}_i^r \times (\bar{\omega}_i^r \times \bar{r}_i)$

Considering expressions for kinematic parameters set forth in Table 5.2, dependences are obtained for the absolute velocities and accelerations of the point O_i *of the i*-link, as well as angular velocities and accelerations of the *i*- link relative to the fixed coordinate system. Expressions given below are obtained considering the type of the kinematic pair formed by the links *i and i-1*. With translational kinematic pair, the following dependences are introduced for the angular velocity and the absolute linear velocity:

$$\bar{\omega}_{\mathrm{i}}=\bar{\omega}_{\mathrm{i-1}};\bar{v}_i = \bar{v}_{i-1} + \bar{\omega}_{i-1} \times \bar{r}_i + \bar{z}_{i-1}\dot{\theta}_{\mathrm{i}}. \quad (5.35)$$

Vector equality for the angular and linear absolute accelerations are as follows:

$$\bar{\varepsilon}_{\mathrm{i}}=\bar{\varepsilon}_{\mathrm{i-1}};\bar{a}_i = \bar{a}_{i-1} + \bar{\varepsilon}_i \times \bar{r}_i \times 2\bar{\omega}_i \times \left(\bar{z}_{i-1}\dot{\theta}_{\mathrm{i}}\right) + \bar{\omega}_i \times (\bar{\omega}_i \times \bar{r}_i) + \bar{z}_{i-1}\ddot{\theta}_{\mathrm{i}}. \quad (5.36)$$

If the *i*- and *i-1*-links form the rotational kinematic pair, the angular velocities and accelerations, as well as linear velocities and accelerations are determined by the following expressions

$$\begin{aligned} &\bar{\omega}_i=\bar{\omega}_{i-1}+\bar{z}_{i-1}\dot{\theta}_{\mathrm{i}};\bar{v}_{\mathrm{i}}=\bar{v}_{\mathrm{i-1}}+\bar{\omega}_i \times \bar{r}_i; \\ &\bar{\varepsilon}_{\mathrm{i}}=\bar{\varepsilon}_{\mathrm{i-1}}+\bar{\omega}_{i-1} \times \left(\bar{z}_{i-1}\dot{\theta}_{\mathrm{i}}\right)+\bar{z}_{i-1}\ddot{\theta}_{\mathrm{i}}; \\ &\bar{a}_i = \bar{a}_{i-1} + \bar{\varepsilon}_i \times \bar{r}_i + \bar{\omega}_i \times (\bar{\omega}_i \times \bar{r}_i). \end{aligned} \quad (5.37)$$

The obtained by the vector method kinematic parameters are applied hereinafter for solving dynamics problem using the D'Alembert principle. It is more convenient to solve these problems in the local coordinate system related to the link, based on the solutions obtained, for example, in the study [46].

5.5.2 Kinematic Parameters in the Link Coordinate System

The output of relations for the kinematic parameters in the local coordinate system related to the i – link, is based on the formulas (5.37). While it uses a part of the kinematic pair matrix that describes the transformation of the systems rotation relative to each other:

$$R_i^{i-1} = \begin{pmatrix} C\theta_i & -C\alpha_i S\theta_i & S\alpha_i S\theta_i \\ S\theta_i & C\alpha_i S\theta_i & -S\alpha_i C\theta_i \\ 0 & S\alpha_i & C\alpha_i \end{pmatrix}. \quad (5.38)$$

It should be noted that the matrix R_i^{i-1} is orthonormal, therefore $\left(R_i^{i-}\right)^{-1} = R_{i-1}^i = \left(R_i^{i-1}\right)^T$. Hence,

$$R_{i-1}^{i} = \begin{pmatrix} C\theta_i & S\theta_i & 0 \\ -C\alpha_i S\theta_i & C\alpha_i S\theta_i & S\alpha_i \\ S\alpha_i S\theta_i & -S\alpha_i C\theta_i & C\alpha_i \end{pmatrix}. \tag{5.39}$$

The matrix R_{i-1}^{i} transforms the three-dimensional vector from the i-1 coordinate system into *i*-system. All the unknown kinematic parameters in the expressions (5.37) are described in the *i-1*-coordinate system. Using the matrix R_{i-1}^{i} the same kinematic parameters can be described in the local *i*-coordinate system. In this case, the formulas (5.37) is transformed with the translational kinematic pair as follows:

$$\begin{aligned}
R_0^i\bar{\omega}_i &= R_{i-1}^{i}\left(R_0^{i-1}\bar{\omega}_{i-1}\right), \\
R_0^i\bar{v}_i &= R_{i-1}^{i}\bar{z}_{i-1}\dot{\theta}_i + \left(R_0^i\bar{\omega}_i\right)\times\left(R_0^i\bar{r}_i\right) + R_{i-1}^{i}\left(R_0^{i-1}\bar{v}_{i-1}\right), \\
R_0^i\bar{\varepsilon}_i &= R_{i-1}^{i}\left(R_0^{i-1}\bar{\varepsilon}_{i-1}\right), \\
R_{i-1}^{i}\bar{a}_i &= R_{i-1}^{i}\left(\bar{z}_0\ddot{\theta}_i + R_0^{i-1}\bar{a}_{i-1}\right) + \left(R_0^i\bar{\varepsilon}_i\right)\times\left(R_0^i\bar{r}_i\right) \\
&\quad + 2\left(R_0^i\bar{\omega}_i\right)\times\left(R_0^i\bar{z}_0\dot{\theta}_i\right) + \left(R_0^i\bar{\omega}_i\right) \\
&\quad \times\left[\left(R_0^i\bar{\omega}_i\right)\times\left(R_0^i\bar{r}_i\right)\right].
\end{aligned} \tag{5.40}$$

If the interconnected links form the rotational kinematic pair, then the expressions for the velocities and accelerations, the angular velocities and accelerations in the local coordinate system will look as follows:

$$\begin{aligned}
R_0^i\bar{\omega}_i &= R_{i-1}^{i}\left(\bar{z}_0\dot{\theta}_i + R_0^{i-1}\bar{\omega}_{i-1}\right), \\
R_0^i\bar{v}_i &= \left(R_0^i\bar{\omega}_i\right)\times\left(R_0^i\bar{r}_i\right) + R_{i-1}^{i}\left(R_0^{i-1}\bar{v}_{i-1}\right), \\
R_0^i\bar{\varepsilon}_i &= R_{i-1}^{i}\left[R_0^{i-1}\bar{\varepsilon}_{i-1} + \bar{z}_0\dot{\theta}_i + \left(R_0^{i-1}\bar{\omega}_{i-1}\right)\times\bar{z}_0\dot{\theta}_i\right], \\
R_{i-1}^{i}\bar{a}_i &= \left(R_0^i\bar{\varepsilon}_i\right)\times\left(R_0^i\bar{r}_i\right) + \left(R_0^i\bar{\omega}_i\right) \\
&\quad \times\left[\left(R_0^i\bar{\omega}_i\right)\times\left(R_0^i\bar{r}_i\right) + R_{i-1}^{i}\times\left(R_0^{i-1}\bar{a}_i\right)\right].
\end{aligned} \tag{5.41}$$

As the result, the unknown vector equation are found; they define different kinematic parameters in the local coordinate system related to the link in the cases, where the interconnected links perform translational or rotational motions in the relative motion.

5.5.3 Kinematic Parameters of the Connecting Link and the Gripper

One of the characteristics of the manipulators with closed kinematic chain, designed according to the method adopted in the present study, is the presence of the *CL* in their design. In this regard, the task to define the kinematic parameters of the *CL* is set.

From previous findings expression setting velocity and acceleration of two points of the connecting link are known. The solution of the set problem begins with identifying the angular velocity of the connecting link, which is considered as a free body. For this purpose, the vector equality based on the theorem of the velocity distribution during the free motion of the body. Having accepted the point **B** as plus, velocity of which is known, vector equality is recorded for determining (known) velocity of the point **A**

$$\bar{v}_A = \bar{v}_B + \bar{\omega}_C \times (\bar{r}_A - \bar{r}_B). \tag{5.42}$$

In the equation, the vectors $\bar{r}_A, \bar{r}_B$ determine the position of the points **A** and **B** relative to the reference coordinate system.

In the manipulators with closed kinematic chains, as well as in the mechanism of the six-axis table discussed in this paper, the connecting link according to its functional features cannot rotate around its own axis, i.e.

$$\bar{\omega}_C \times (\bar{r}_A - \bar{r}_B)= 0. \tag{5.43}$$

From the vector analysis, it is known that when the scalar product is given according to (5.42), then the vector product

$$(\bar{r}_A - \bar{r}_B) \times \bar{\omega}_C=\bar{v}_A - \bar{v}_B$$

has the only inverse solution

$$\bar{\omega}_C = [(\bar{v}_B - \bar{v}_A) \times (\bar{r}_A - \bar{r}_B)]/l_{AB}^2. \tag{5.44}$$

where l_{AB} is the length of the connecting link.

The projections of the vector Eq. (5.44) on the axis X_o, Y_o, Z_o of the fixes coordinate system provide expressions for the elements of the angular velocity ω_{cx}, ω_{cy}, ω_{cz} of the connecting link **AB**

$$\begin{aligned}
\omega_{cx} &= \left[(v_{BY} - v_{AY})(z_B - z_A) - (v_{Bz} - v_{Az})(y_B - y_A)\right]/l_{AB}^2,\\
\omega_{cy} &= \left[(v_{Bz} - v_{Az})(x_B - x_A) - (v_{Bx} - v_{Ax})(z_B - z_A)\right]/l_{AB}^2,\\
\omega_{cz} &= \left[(v_{Bx} - v_{Ax})(y_B - y_A) - \left(v_{By} - v_{Ay}\right)(x_B - x_A)\right]/l_{AB}^2.
\end{aligned} \tag{5.45}$$

Vector of the angular velocity of the connecting link equals to

$$\bar{\omega}_c = \omega_{cx}\bar{i}_0 + \omega_{cy}\bar{j}_0 + \omega_{cz}\bar{k}_0, \tag{5.46}$$

where $\bar{i}_0$, $\bar{j}_0$, $\bar{k}_0$ are cross headings of the fixed coordinate system.

In the case where the connecting link can rotate around its own axis, i.e., if the condition (5.43) is not satisfied, it is possible at the preliminary stage of calculations to set the velocities v_1, v_2, v_3 to three non-coplanar points of the connecting link. Then, the angular velocity of the connecting link is determined according to [43]

$$\bar{\omega}_c = 2(\bar{v}_1 \times \bar{v}_2 + \bar{v}_2 \times \bar{v}_3 + \bar{v}_3 \times \bar{v}_1)/[\bar{v}_1(\bar{r}_2 - \bar{r}_3) + \bar{v}_2(\bar{r}_3 - \bar{r}_1) + \bar{v}_3(\bar{r}_1 - \bar{r}_3)]. \tag{5.47}$$

where $\bar{r}_1, \bar{r}_2, \bar{r}_3$ are radius vectors defining the position of the non-coplanar points 1, 2, 3 relative to the reference coordinate system.

Now according to the formula (5.42), velocity of any point of the connecting link can be determined, if the plus is accepted as the point ***A*** or ***B***.

The angular velocity of the gripper that has its own angular rotation velocity of $\bar{\omega}^*_{cx}$, is defined as the angular velocity of the body during complex movement

$$\bar{\omega}_{cx} = \bar{\omega}_c + \bar{\omega}^*_{cx}.$$

Velocity of the gripper center equals to the velocity of the point of the connecting link, which coincides with the center of the gripper, according to (5.42) it is

$$\bar{v}_{cx} = \bar{v}_A + \bar{\omega}_{cx} \times (\bar{\mathrm{r}}_c - \bar{r}_A). \tag{5.48}$$

Sometimes, it is necessary to determine the velocity of an arbitrarily chosen point ***M*** of the gripper, position of which relative to the center of the gripper is determined by the radius vector $\bar{r}^*_{cx}$. This velocity is determined according to the velocity addition theorem during the complex motion of the body

$$\bar{v}_M = \bar{v}_{cx} + \bar{\omega}^*_{cx} \times \bar{\mathrm{r}}^*_{\mathrm{cx}}. \tag{5.49}$$

In this formula $\bar{\omega}^*_{cx} = \dot{\phi}_6 \bar{z}_6$, there is the angular velocity of the gripper rotation.

The projections of the angular acceleration of the connecting link are determined by differentiating the expression for the projections of the angular velocity (5.45)

$$\begin{aligned}
\varepsilon_{cx} &= \left[\left(a_{By} - a_{Ay}\right)(z_B - z_A) - (a_{\mathrm{B}z} - a_{Az})(y_B - y_A)\right]/l^2_{AB},\\
\varepsilon_{cy} &= \left[(a_{Bz} - a_{Az})(x_B - x_A) - (a_{\mathrm{B}x} - a_{Ax})(z_B - z_A)\right]/l^2_{AB},\\
\varepsilon_{cz} &= \left[(a_{Bx} - a_{Ax})(y_B - y_A) - \left(a_{\mathrm{B}y} - a_{Ay}\right)(x_B - x_A)\right]/l^2_{AB}.
\end{aligned} \tag{5.50}$$

Acceleration of the point ***C*** of the connecting link is found by the theorem of the acceleration distribution for the free motion of the solid body

$$\bar{a}_{cx} = \bar{a}_A + \bar{\varepsilon}_c \times (\bar{r}_c - \bar{r}_A) + \bar{\omega}_c \times [\omega_c \times (\bar{\mathrm{r}}_c - \bar{\mathrm{r}}_A)]. \tag{5.51}$$

In this formula, $\bar{a}_A$ is the known acceleration of the point ***B*** of the connecting link. While the point ***B*** is accepted as plus. The angular acceleration of the connecting link equal to

$$\bar{\varepsilon}_c = \varepsilon_{cx}\bar{i}_0 + \varepsilon_{cy}\bar{j}_0 + \varepsilon_{cz}\bar{k}_0.$$

The angular acceleration of the gripper during the complex rotational motion is defined by the equation

$$\bar{\varepsilon}_{cx} = \bar{\varepsilon}_c + \ddot{\phi}_6\bar{z}_6 + \bar{\omega}_c \times \left(\dot{\phi}_6\bar{z}_6\right). \tag{5.52}$$

According to the velocity addition theorem for complex motion of a point, the acceleration of an arbitrary point belonging to the gripper is found as follows:

$$\bar{a}_{cx} = \bar{a}_c + \bar{e}_{cx} \times \bar{r}^*_{cx} + \bar{w}_{cx} \times \left(\bar{w}_{cx} \times \bar{r}^*_{cx}\right).$$

As the result, using the vector method, all the basic kinematic parameters are found for the connecting link and the gripper.

For power analysis method used hereinafter, it is needed to consider the dependence of the kinematic parameters of the connection link and the gripper in the local coordinate system related to the connecting link. To obtain these dependences, the rotation matrix (3 × 3) matrix from the the manipulator matrix (5.1) is used

$$R = \begin{pmatrix} Ctg\gamma \mathrm{Cos}\alpha & Ctg\gamma \mathrm{Cos}\beta & \mathrm{Sin}\gamma \\ -\mathrm{Cos}\beta\mathrm{Cos}2\gamma/\mathrm{Sin}\gamma & \mathrm{Cos}\alpha\mathrm{Cos}2\gamma/\mathrm{Sin}\gamma & 0 \\ \mathrm{Cos}\alpha & \mathrm{Cos}\beta & \mathrm{Cos}\gamma \\ 0 & 0 & 0 \end{pmatrix}. \tag{5.54}$$

For example, similarly to (5.51), acceleration of the mass center of the connecting link is equal to

$$\bar{a}_{cs} = \bar{a}_A + \bar{e} \times (\bar{r}_{cs} - \bar{r}_A) + \bar{w}_C \times [\bar{w}_C \times (\bar{r}_{cs} - \bar{r}_A)], \tag{5.55}$$

where $\bar{r}_{cs}$ is a radius vector defining the position of the mass center of the connecting link relative to the reference coordinate system.

In the local coordinate system $BX_cY_cZ_c$, acceleration a^c_{cs} of the mass center of the connecting link with the gripper is determined by the dependence

$$\bar{a}^c_{cs} = R^{-1}\bar{a}_{cs}. \tag{5.56}$$

where $R^{-1}$is the matrix that is inverse for the matrix R (5.54).This expression is used in the force calculation of the connecting link with the gripper.

5.5.4 Homogeneous Matrix Method for Determining Speeds and Accelerations

When studying dynamics of the manipulators using Lagrange's equations of the 2nd kind, representations of the kinematic parameters are used with the help of the homogeneous transformation matrices. The expressions for the velocities and accelerations of the manipulator links are obtained by using these matrices. Velocity of the point of the i-link with radius vector $\bar{r}_i$ relative to the fixed coordinate system is determined by differentiating the matrix equality (5.9)

$$v_i = \frac{d}{dt}(T_i, r_i) = \left[\sum_{j=1}^{i} \frac{dT_i}{dq_j} q_j\right] r_i. \tag{5.57}$$

where $v_i = (v_{xi}, v_{yi}, v_{zi}, 0)^T$ -is extended column vector made of projections of the velocity of the point of the i-link on the axis of the fixed coordinate system; $r_i = (x_i, y_i, z_i, 1)^T$ is extended column vector, elements of which are coordinates of the point belonging to the *i*-th link; $q_j - j$ is generalized coordinate ($j = 1, \ldots, W$); W is the manipulator's degree of freedom.

It is known that the transformation matrix equals to the product of the matrices of the kinematic pairs: $T_i = A_1 A_2 \ldots A_{n-1} A_n$, where n is the number of kinematic pairs. Given [47] with $j \leq i$

$$U_{ij} = \frac{dT_i}{dq_j} = A_1 A_2 \ldots \Omega_j A_j \ldots A_i, \tag{5.58}$$

And with $j > i$ $U_{ij} = 0$, the expression for the velocity of the i-th point finally has the following form:

$$v_i = \left[\sum_{j=1}^{i} U_{ij} q_i\right] r_i. \tag{5.59}$$

Matrices A and Ω were determined earlier.

Using the extended homogeneous transformation matrices T_i, by further differentiation, it is possible to obtain the expression for the acceleration of the point belonging to the i-th link, and namely:

$$a_i = \ddot{T}_i r_i, \tag{5.60}$$

where $a_i = (a_{xi}, a_{yi}, a_{zi}, 0)^T$ is the extended column vector made of projections of the velocity of the point of the i-link on the axis of the fixed coordinate system. While, the second derivative of the transformation matrix is

$$\ddot{T}_i = \sum_{j=1}^{i}\sum_{k=1}^{i} V_{ijk}\dot{q}_j\dot{q}_k + \sum_{j=1}^{i} U_{ij}\ddot{q}_j. \tag{5.61}$$

Here, for convenience, the auxiliary matrix is introduced

$$V_{ijk} = \frac{dU_{ij}}{dq_k} = V_{ikj}. \tag{5.62}$$

This matrix is expressed by the product of the homogeneous transformation matrices just like (5.58) taking into account the expressions for $\Omega_{i.}$

Thus, with the known values of the generalized velocities and accelerations, and the predetermined matrices of the kinematic pairs, using the formulas (5.58) and (5.59), velocity and acceleration can be determined for any point of the *i*-th link, which is fixed relative to the *i*-the link.

5.5.5 *Generalized Velocities and Accelerations*

Here dependences are derived for the generalized velocities and accelerations. The results obtained will be used in determining the kinematic parameters necessary for the power analysis of the manipulators. Below are solutions for two problems of calculating the generalized velocities and accelerations.

The first problem: given—velocity and acceleration of the gripper center, the angular velocity and acceleration of the connecting link; it is required to determine the generalized velocities and accelerations of the CFG links.

Second problem: given—values of the generalized coordinates obtained by solving inverse kinematics problem; it is required to find the generalized velocities and accelerations of the manipulator.

The choice of a solution depends on what parameters are assumed to be given. If the values of the orientation angles of the connecting link at different points at equal intervals of time are given, then calculating the derivatives of α and β by the formulas of the finite differences [60], the following result can be obtained

$$\begin{aligned} \dot{\alpha}_i &= (\alpha_{i+1} - \alpha_{i-1})/2\tau, \ddot{\alpha}_i = (\alpha_{i+1} + \alpha_{i-1} - 2\alpha_i)/\tau^2, \\ \dot{\beta}_i &= (\beta_{i+1} - \beta_{i-1})/2\tau, \ddot{\beta}_i = (\beta_{i+1} - \beta_{i-1})/2\tau^2. \end{aligned} \tag{5.63}$$

Here the subscript of the variables indicates the number of the gripper position, in which this variable is found. For example, $\alpha_{i\text{-}1}$ represents the inclination to the axis OX of the connecting link in the *i-1* position, while *i* = *1*,...,*n*, where n is the number of intervals, in which the path of the gripper center is divided. In the Eqs. (5.62), τ denotes the time of motion within the same interval.

The angular velocity of the connecting link equals

$$\omega_{ci} = \sqrt{\dot{\alpha}_i^2 + \dot{\beta}_i^2}. \tag{5.64}$$

Similarly, the angular acceleration of the connecting link is found

$$\varepsilon_{ci} = \sqrt{\ddot{\alpha}_i^2 + \ddot{\beta}_i^2}. \tag{5.65}$$

With the known values of the velocity, acceleration of the gripper center and the angular velocity, angular acceleration of the connecting link, by the formulas (5.48, 5.51) of the velocity and acceleration distribution, velocity and acceleration of the output point of each CFG (v_{ai}, v_{θ}, a_{ai}, $a_{\theta i}$) can be found. Since the homogeneous transformation matrix for each of the CFG (5.2–5.7) is determined, it is easy to find the projections of the velocity and acceleration of the output point of the CFG on the axis of the reference coordinate system.

By double differentiation of the Eqs. (5.10–5.15), connecting generalized coordinates of the CFG with the coordinates of the output point set the relation between the generalized velocities, accelerations and velocities, accelerations of the output point of the CFG. Thus, the generalized velocities and accelerations are determined at each position of the manipulator.

When solving the second problem, according to the results of the inverse kinematic problem solution, *j*-generalized coordinates of the manipulator are found at each *i*-th position (q_{ji}).

The drive operation cycle is divided into three intervals corresponding to the acceleration mode, the set steady motion and deceleration. It is assumed that at the cycle start and end, the velocity is zero.

Maximal velocity $v_j^{\max}$ and acceleration $a_j^{\max}$ developed by the drive are deemed to be predetermined. These values are known from the technical capabilities of the drive, as well as from its operating restrictions.

Let us assume that the generalized acceleration varies sinusoidally [48].

Then the acceleration time is found by the formula

$$t_{pj} = \pi v_j^{\max} / 2a_j^{\max}. \tag{5.66}$$

The total time of the gripper motion (one cycle period) is equal to

$$t_{nj} = \frac{q_{jn} - q_{jo}}{v_j^{\max}} + 2t_{pj}. \tag{5.67}$$

During the acceleration with $0 \leq t_i < t_{pj}$, the generalized *j*-th velocity and acceleration vary according to the following dependences:

$$\begin{aligned} \dot{q}_{ji} &= a_j^{\max} t_{pj} (1 - \mathrm{Cos}\pi i t_i)/\pi, \\ \ddot{q}_{ji} &= a_j^{\max} \mathrm{Sin}\pi t_i / t_{pj}. \end{aligned} \tag{5.68}$$

With $t_{pj} \leq t_i < t_{nj} - t_{pj}$ corresponding to the steady motion mode, the generalized j-th velocity and acceleration are determined by formulas

$$\dot{q}_{ji} = v_j^{\max} + 2a_j t_{pi}/\pi, \ddot{q}_{ji} = 0 \tag{5.69}$$

In case of deceleration ($t_{nj}-t_{pj} \leq t_i < t_{nj}$), the generalized j-th velocity and acceleration are determined by the equations

$$\begin{aligned} \dot{q}_{ji} &= \frac{a_j^{\max} t_{pj}}{p}\left[1 + Cos\frac{p}{t_{pj}}(t_i - t_{\bullet j} + t_{pj})\right], \\ \ddot{q}_{ji} &= -a_j^{\max} Sin\left[p(t_i - t_{\bullet j} + t_{pj})/t_{pj}\right]. \end{aligned} \tag{5.70}$$

If calculations are made at regular intervals, the duration of the interval equals $\tau_\iota = t_{nj}/n$, a $t_i = \iota\tau_\iota$.

These dependences determining the kinematic parameters are used in the power analysis of the multi-degree-of-freedom mechanisms of the robotic manipulators at the stage of design and selection of drives.

Chapter 6
Kinematics of Multiloop Parallel Manipulators

Abstract The kinematic analysis of the MPM uses the redundant parameters method that allow obtaining analytical description of links position in any mechanisms and manipulators, as well as obtaining more simple final matrix expression (with smaller number of terms). Using this method, the direct and inverse kinematics problem of the SHOLKOR parallel manipulator is solved.

Keywords Direct and inverse kinematics problem · Redundant parameters method · SHOLKOR kinematics

6.1 The Features Conditioned by the Design of the Multiloop Parallel Manipulator

The review of the platform manipulators conducted by authors [1–3], shows that for the manipulators of the platform type, explicitly there is no solution for the direct and inverse position problems. In such platform manipulators, troubles appear in creating control systems and building control commands.

Discussed below multiloop parallel manipulator of the SHOLKOR robot, due to the design organization (block diagram), allows getting solutions for the direct and inverse position problem, and thus provides an opportunity to use all the advantages of the automatic control.

SHOLKOR MPM has a range of features conditioned by its construction (design) and playing an important role in the kinematic research in the future and in the control.

Firstly, one platform has 6 degrees of freedom relative to the other platform. While 6 drives located on the connecting links moving independently from each other, ensure these degrees of freedom.

Secondly, movement of the platform (Fig. 6.1) can be performed by positioning of some straight line B_2C_2 in the space, which is implemented by transferring the node C_2 to the set point of the space, by changing the length of the connecting links 3, 6, 7; and then node B_2 is positioned by changing the length of the connecting links

K. S. Sholanov, *Parallel Manipulators of Robots*, Mechanisms and Machine Science 92,
https://doi.org/10.1007/978-3-030-56073-7_6

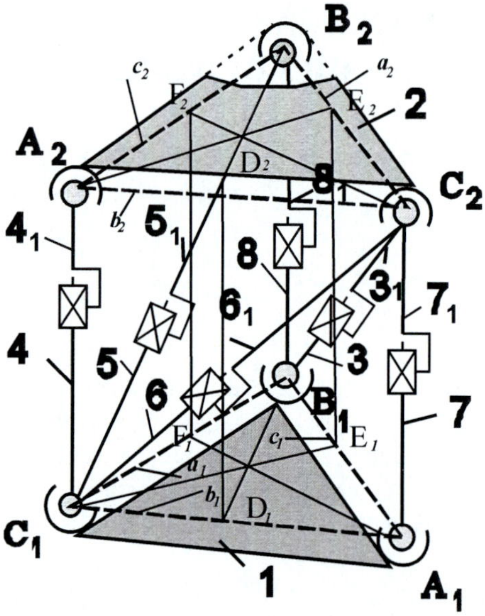

Fig. 6.1 Multiloop manipulator mechanism

5, 8. Positioning of the upper platform relative to the bottom platform is completed by the rotation around a straight line B_2C_2, change of the node A_2 position.

Thirdly, a prism that is represented by the MPM SHOLKOR, can always be broken down into three pyramids. For this reason, the analysis of the manipulator kinematics can consider the motion of the nodes of individual pyramids.

From the above, it follows that the motion of the platform 2 relative to the platform 1 can be divided into three consecutive phases: movement of the node C_2, and then nodes B_2 and A_2. According to this sequence below, the kinematic analysis of this manipulator is performed.

For the kinematic analysis, the geometric dimensions, as well as the starting position of the manipulator with the configuration in the form of rectangular prism (Fig. 6.1) are deemed predetermined, while the lengths of the connecting links correspond to the sum of their geometrical lengths and a half of their stroke. Herein, the stroke $h_{i\max}$ is the remainder of the maximal length of the i-connecting link and its minimal geometric dimension l_{0i}.

Subsequently, let us agree to determine the length of the connecting links at their current position as follows:

$$B_1C_2 = l_3 + h_3;\ C_1A_2 = l_4 + h_4;\ C_1B_2 = l_5 + h_5;\ C_1C_2 = l_6 + h_6;\ A_1C_2 = l_7 + h_7;\ B_1B_2 = l_8 + h_8.$$

where $l_3 - l_8$ are lengths of the connecting links in the previous position, for example, in its initial position $l_3 = l_{03} + h3\max/2$; l_{0i} is the minimal value of the lengths of the connecting links; $h_3 - h_8$ are positive or negative displacements, representing the displacement increment (according to the sensor readings). These movements are performed by the drives of the respective connecting links. On the bottom 1 and upper 2 platform, triangles $A_1B_1C_1$ and $A_2B_2C_2$ are outstand, vertices of which are node points. Hereinafter, it is assumed that the triangles of the upper and bottom platform are regular, so is believed that $a_1 = в_1 = c_1 = a_2 = в_2 = c_2 = a$. Also, to simplify the calculations, let us assume that the minimum dimensions of the edges are $l_{04} = l_{07} = l_{08} = ka$ *(k is a coefficient chosen due to design reasons).* Diagonal lines of the rectangles of the prism faces are equal, hence: the lengths of the diagonal edges equal: $l_{03} = l_{05} = l_{06} = a\sqrt{1+k^2}$.

It should be noted that the configuration of the faces and the condition of links continuity set limitations for the maximum displacements permitted by separate drives.

6.2 Solution a Direct Position Problem of the Platform for the Multiloop Manipulator

To determine the position of the node C_2 of the upper platform, let us consider the pyramid $C_2A_1B_1C_1$ (Fig. 6.2). First, the axes of the reference coordinate system $C_1X_1Y_1Z_1$ are selected. The center of the coordinate system $C_1X_1Y_1Z_1$ related to the platform 1, is chosen in the node C_1.vAxis C_1X_1 is directed along the side A_1C_1 of the triangle $A_1B_1C_1$ away from C to A. Axis C_1Z_1 is directed perpendicularly to the plane of the triangle $A_1B_1C_1$. When selecting the reference coordinate system and moving coordinate systems related to the connecting link, there is a certain choice, but the coordinate system must be chosen so that the angles used in the transformation are related to the prism configuration, i.e., so that they can be identified for the current configuration. It is also necessary that the chosen transformation parameters are related to the constant geometric dimensions and the unknown variables of the manipulator values.

Fig. 6.2 Basic pyramid and the transformation parameters by the link 6

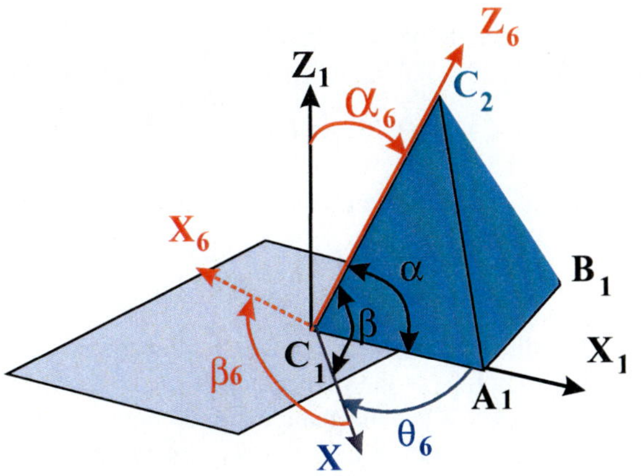

Further, for the kinematic analysis, the method of 8 parameters is used (paragraph 4.2–4.4) [45]. Node C_2 as a point in the space has 3 degrees of freedom, the node position depends on the movement h_3, h_6, h_7 of the connecting links 3,6,7. To get the dependencies for determining the position of the node C_2, let us perform transformations of the moving coordinate systems through three connecting chains along the paths C_I–C_2, C_I–B_I–C_2, C_I–A_I–C_2 till coincidence with the reference coordinate system. At the same time the centers of the moving coordinate systems related to the connecting links, are selected in the centers of the spheres, axes Z are directed along the axes of the connecting links, axes X are directed perpendicular to the axes Z in the plane of the side face. Let us consider, first, the transformation of the moving coordinate system $C_IX_6Y_6Z_6$ related to the connecting link 6, into the reference coordinate system $C_IX_IY_IZ_I$. This transformation, as shown in Fig. 6.2, is implemented by a range of successive movements of the system $C_IX_IY_IZ_I$ until complete coincidence with the coordinate system $C_IX_6Y_6Z_6$. The first movement of the transformation is the rotation of the system $O_IX_IY_IZ_I$ for the angle of θ_6 around the axis C_IZ_I until the axis C_IX_I is perpendicular to the axis C_IZ_6 ($\beta = \pi / 2$). The new position of the axis C_IX_I is denoted as C_IX. Surface $C_2A_IC_I$ (side face $A_IC_IA_2C_2$ of the prism in Fig. 6.2) forms the dihedral angle $\angle C$ with the plane $C_IX_IY_I$ of the base. According to the theorem of cosines, for the trihedral angle with the vertex C_I there is

$$Cos\beta = Cos\alpha \cdot Cos\theta_6 + Sin\alpha \cdot Sin\theta_6 \cdot Cos(\pi - C).$$

Since $\beta = \pi/2$, after simplification the following equality is obtained

$$Ctg\theta_6 = tg\alpha \cdot CosC. \tag{6.1}$$

where α is the angle at the base of the triangle $A_IC_I\,C_2$. In order to determine the angle α, let us perform the following actions: define the semiperimeter of the triangle $A_IC_I\,C_2$

$$p = (a + l_7 + h_7 + l_6 + h_6)/2; \tag{6.2}$$

define the length of the height lowered from the vertex C_2

$$H = \frac{2}{a}|\sqrt{p(p-a)(p-l_7-h_7)(p-l_6-h_6)}|. \tag{6.3}$$

Hence, the angle α at the base of the triangle $C_2A_IC_I$ equals

$$\alpha = arc\,\mathrm{Cos}(H/(l_6 + h_6)). \tag{6.4}$$

The dihedral angle $\angle C$ equals the linear angle between D_ID_2 and height of the base D_IB_I.D_ID_2 is the perpendicular recovered from the middle of the side of the triangle A_IC_I. When the manipulator configuration is varied, coordinates of the point D_2 change. In determining the coordinates of the point D_2, 3 cases are discussed depending on the value of the angle α (Fig. 6.3).

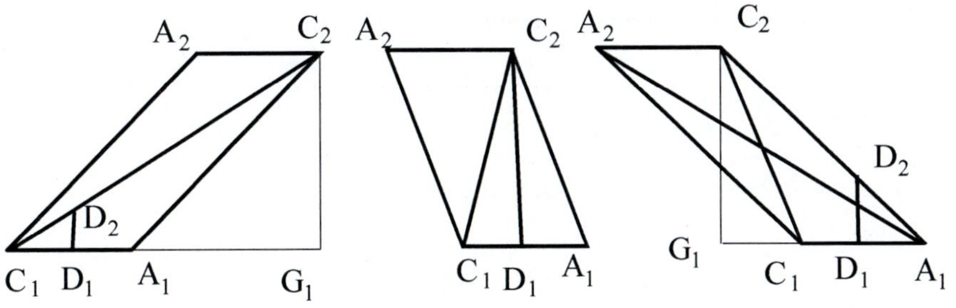

Fig. 6.3 Possible change of the edge configuration

1 case, when $(l_6 + h_6)Cos\alpha > a/2$ (Fig. 6.3a).

In this case, to determine the ratio, in which the point D_2 divides the line A_2C_2, from the vertex C_2 a perpendicular is lowered to the side A_1C_1 and the point is denoted as G_1. The base is defined by the expression $C_1G_1 = (l_6 + h_6)\text{Cos}\,\alpha$. Two perpendiculars D_1D_2 and C_2G_1 divide the lines A_1C_1 and C_1C_2 on equal parts, and namely:

$$\lambda = \frac{C_1G_1}{C_1C_2} = \frac{C_1D_1}{C_1D_2} = \text{Cos}\alpha.$$

Hence,

$$C_1D_2 = r = \frac{a}{2\,\text{Cos}\,\alpha},\ D_1D_2 = f = \frac{a}{2}tg\alpha. \tag{6.5}$$

Let us now consider the triangle $C_1C_2B_1$ (Fig. 6.4a) and find D_2B_1. From the point B_1, a perpendicular is lowered to the side C_1C_2 and denote the perpendicular base with K.

The semiperimeter of the triangle is

$$p = (a + l_3 + h_3 + l_6 + h_6)/2. \tag{6.6}$$

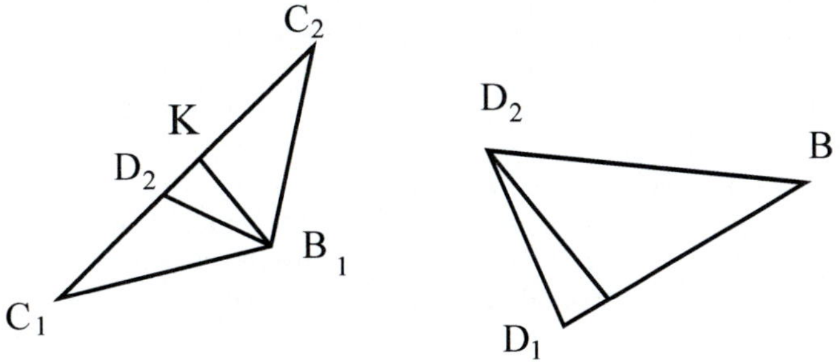

Fig. 6.4 Auxiliary constructions

Height B_1K is identified by the equation

$$H = \frac{2}{l_6 + h_6} |\sqrt{p(p-a)(p-l_3-h_3)(p-l_6-h_6)}| \tag{6.7}$$

The angle at the base of the triangle $C_1C_2B_1$

$$Sin\chi = \frac{H}{a},$$

side C_1K equals

$$l_k = H \cdot tg\chi = \frac{H^2}{|\sqrt{a^2 - H^2}|}. \tag{6.8}$$

Let us find the section KD_2

$$KD_2 = m = |r - l_k|$$

and from the right-angled triangle KD_2B_1, D_2B_1 *is found*

$$D_2B_1 = s = \left|\sqrt{(r-l_k)^2 + H^2}\right| \tag{6.9}$$

Let us consider the triangle $D_1D_2B_1$, in which it is needed to find semiperimeter and the height lowered from the vertex D_2:

$$p = (a\sqrt{3}/2 + s + f)/2, \tag{6.10}$$

$$H = \frac{4}{a\sqrt{3}} |\sqrt{p(p - a\sqrt{3}/2)(p - s_3)(p - f)}|. \tag{6.11}$$

Linear angle that is equal dihedral angle $\angle C$ is defined by the expression

$$\text{Cos}(C) = \frac{H}{f}. \tag{6.12}$$

In the initial (starting) position, the value $\angle C = \pi/2$, of the displacement in drives is zero, i.e. $h_6 = h_7 = 0$.

2 case, when $(l_6 + h_6)\,\text{Cos}\alpha = a/2$ (Fig. 6.3b).

In this case, the perpendicular recovered from the point D_2, passes through the node C_2. In this case, the triangle $C_2\ D_2B_1$ is discussed, sides of which are known. Its semiperimeter and height are determined by the equations

$$p = (a\sqrt{3}/2 + \left|\sqrt{(l_6 + h_6)^2 - a^2/4}\right| + l_3 + h_3)/2,$$
$$H = \tfrac{4}{a\sqrt{3}} |\sqrt{p(p - a\sqrt{3}/2)(p - \left|\sqrt{(l_6 + h_6)^2 - a^2/4}\right|)(p - l_3 - h_3)}|. \quad (6.13)$$

Linear angle that is equal dihedral angle $\angle C$ is defined by the expression

$$\mathrm{Cos}(C) = \frac{H}{\left|\sqrt{(l_6 + h_6)^2 - a^2/4}\right|}.$$

3 case, when $(l_6 + h_6)\,\mathrm{Cos}\alpha < a/2$ (Fig. 6.3c).

In this case, the perpendicular recovered from the point D_1, crosses the edge A_1C_1 at the point D_2. From the triangle $A_1C_1C_2$, it can be found that

$$C_1G_1 = |(l_6 + h_6)\mathrm{Cos}\,\alpha|.$$

Let us make-up a ratio

$$\lambda = \frac{A_1D_2}{A_1D_1} = \frac{l_7 + h_7}{a + |(l_6 + h_6)\,\mathrm{Cos}\,\alpha|},$$

hence

$$A_1D_2 = v = \frac{a(l_7 + h_7)}{a + |(l_6 + h_6)Cos\alpha|},\; D_1D_2 = f = \left|\sqrt{v^2 - a^{2/4}}\right|.$$

Let us consider the triangle $A_1D_2B_1$ (Fig. 6.5a) to find the length of the section D_2B_1. Since the sides of the triangle $A_1D_2B_1$ are known, the solution sequence given above is used, in detail: the semiperimeter and height of the triangle should be found $p = (a + l_7 + h_7 + l_3 + h_3)/2, H = \frac{2}{l_7+h_7}|\sqrt{p(p - a)(p - l_7 - h_7)(p - l_3 - h_3)}|$.

The angle at the base of the triangle

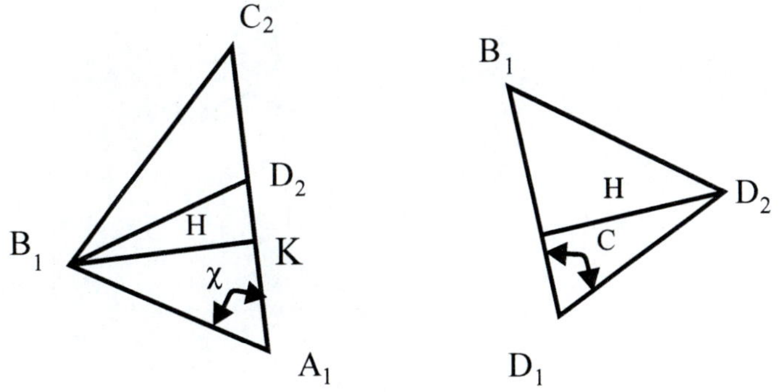

Fig. 6.5 Auxiliary triangles

$\operatorname{Sin}\chi = \frac{H}{a}$, side $A_1K = l_k = H \cdot tg\chi = \frac{H^2}{|\sqrt{a^2-H^2}|}$.,
From the triangle B_1KD_2, the unknown value of D_2B_1 *is found*

$$s = \left|\sqrt{H^2 + (v - l_k)^2}\right|.$$

Now, let us consider the triangle $D_1D_2B_1$, sides of which are known, and determine the desired linear angle equal to the dihedral angle $\angle C$ in the given above sequence
$p = (a\sqrt{3}/2 + f + s)/2$, $H = \frac{4}{a\sqrt{3}}|\sqrt{p(p - a\sqrt{3}/2)(p - f)(p - s)}|$, $\operatorname{Sin}C = \frac{H}{f}$.
The second motion is rotation for the angle of α_6 around the axis C_1X until the axis C_1Z_1 and axis C_1Z_6 coincide. This is the angle that needed to be found. The next motion is rotation around the axis C_1Z_1 for the angle of β_6 until axis C_1X and C_1X_6 *coincide*. This angle does not affect the configuration of the manipulator, so there is no need to calculate this angle. Now let us determine the position of the node C_2 by the transformation along the chain C_1–B_1–C_2. For this moving coordinate system $B_1X_3Y_3Z_3$ related to the connecting link 3 (Fig. 6.6) is selected. The axis B_1Z_3 is directed along the edge B_1C_2, axis B_1X_3 is placed perpendicular to the plane of the face $B_1C_1C_2$. To determine the parameters, when aligning the reference system $C_1X_1Y_1Z_1$ with the coordinate system $B_1X_3Y_3Z_3$, the following actions are performed: rotate the coordinate system $C_1X_1Y_1Z_1$ around the axis C_1Z_1 for the angle of $\theta = \pi/3$, until the axis C_1X is directed along the side B_1C_1 and takes the position of C_1X; then perform parallel transfer of the system from the point C_1 to B_1 along the axis B_1X for the value of a, the new position of the coordinate system is to be denoted as B_1XYZ, The next motion is the rotation of the system B_1XYZ relative the axis B_1Z; for the angle of θ_3, until the axis B_1X is located in the plane of the base perpendicular to the face B_1C_2 ($\beta = \pi/2$) and takes the position of B_1X_{03}. Then rotation is performed relative to the axis B_1X_{03} for the angle of α_3 until the axis B_1Z coincides with the axis B_1Z_3; to align the axes X, rotation is performed

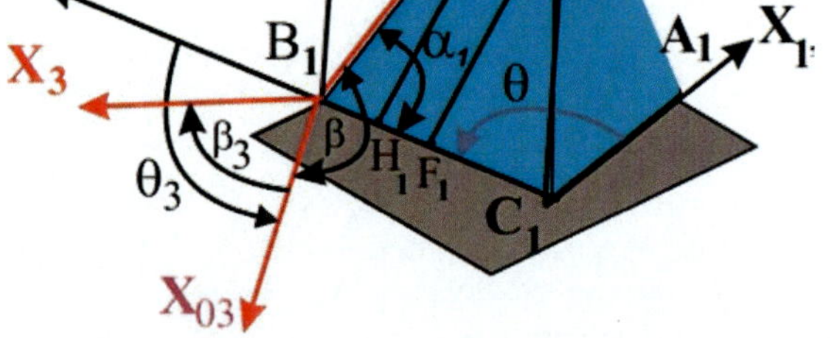

Fig. 6.6 Parameters in the transformation thorough the Link 3

around the axis B_IZ_3 for the angle of β_3. Further, some transformation parameters should be expressed via the known values. According to the theorem of cosines, for the trihedral angle with the vertex B_I there is

$$\mathrm{Cos}\beta = \mathrm{Cos}\alpha_1 \cdot \mathrm{Cos}(\pi - \theta_3) + \mathrm{Sin}\alpha_1 \cdot \mathrm{Sin}(\pi - \theta_3) \cdot \mathrm{Cos}(\pi - B).$$

where $\beta = \pi/2$, is dihedral angle $\angle B$ formed by the plane $C_2B_IC_I$ and the plane $C_IX_IY_I$ of the base. After simplification, the following is obtained

$$Ctg\theta_3 = -tg\alpha_1 \mathrm{Cos}B.$$

Angle α_1 at the base of the face $C_2B_IC_I$ is found similarly to the obtained above angle α. Semiperimeter of the triangle $C_2B_IC_I$

$$p = (a + h_8 + l_8 + h_3 + l_3)/2; \tag{6.14}$$

the length of the perpendicular lowered from the vertex C_2 to the base B_IC_I

$$H = \frac{2}{a}|\sqrt{p(p - a)(p - h_3 - l_3)(p - h_6 - l_6}|. \tag{6.15}$$

The angle α_1 at the triangle base equals

$$\alpha_1 = \arcsin(H/(l_3 + h_3)). \tag{6.16}$$

The dihedral angle $\angle B$ is found by the method used above. Let H_2 be the point of crossing the perpendicular recovered from the middle F_I of the side B_IC_I. To find the coordinates of the point H_2, let us draw the perpendicular down from the point C_2 to the base B_IC_I and determine the projection of the point to the line B_IC_I.

In the initial position of the manipulator, position of the angle C_2 is denoted by the point C_2, however, depending on the configuration, various positions of the angle C_2 (Fig. 6.7) are possible:

1. The angle $\alpha_1 = \pi/2$, while the angle C_2 takes the position of C_{21};

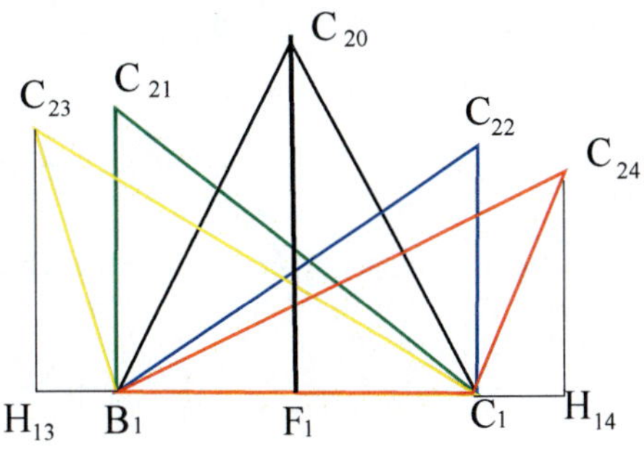

Fig. 6.7 Position of the angle C_2 of the face $B_IC_IC_2$

2. The angle $\alpha_1 < \pi/2$, but not the line 3 projection equals $(l_3 + h_3)Cos\alpha_1 = a$, while the angle C_2 takes the position of C_{22};
3. In this case, when $\alpha_1 < \pi/2$ and $a/2 < (l_3 + h_3)Cos\alpha_1 < a$, the angle C_2 takes the position of C_{24};
4. In the case, when $\alpha_1 < \pi/2$ and $(l_3 + h_3)Cos\alpha_1 = a/2$ the angle C_2 takes the position of C_{20};
5. If the angle $\alpha_1 > \pi/2$, then the angle C_2 takes the position of C_{23}

Let us find the ratio λ, where the point H_2 divides the sides B_1C_2or C_1C_2 in the above cases. In case 1 − $\lambda = 1$, i.e., the point H_2 divides the side C_1C_{21} into halves, wherein the perpendicular length equals to

$$F_1H_2 = f = \left|\sqrt{(l_6 + h_6)^2 - a^2/4}\right|,\ C_1H_2 = r = (l_6 + h_6)/2. \tag{6.17}$$

From the triangle $F_1H_2A_1$, the following is seen

$$A_1H_2 = s = \left|\sqrt{3a^2/4 + f^2}\right|. \tag{6.18}$$

Now all the sides of the triangle $F_1H_2A_1$ are known, therefore, using its semiperimeter and height, the unknown linear angle equal to the dihedral angle $\angle B$ can be found.

Semiperimeter equals to

$$p = (\sqrt{3}a/2 + f + s)/2, \tag{6.19}$$

height

$$H = \frac{4}{a\sqrt{3}}|\sqrt{p(p - \sqrt{3}a/2)(p - f)(p - s)}|. \tag{6.20}$$

The unknown angle, in this case, equals to

$$\mathrm{Sin}B = \frac{H}{f}. \tag{6.21}$$

In case 2—where, there also $\lambda = 1$, i.e., the point H_2 divides now the side B_1C_{22} into halves

$$F_1H_2 = f = \left|\sqrt{(l_3 + h_3)^2/4 - a^2/4}\right|,\ B_1H_2 = r = (l_3 + h_3)/2.$$

Using the triangle $A_1B_1C_2$, sides of which are known, the median A_1H_2 *length is determined*

$$A_1H_2 = s = \frac{1}{2}\left|\sqrt{2a^2 + 2(l_7 + h_7)^2 - (l_3 + h_3)^2}.\right|$$

Now, all the sides of the triangle $A_1H_2F_1$ are known. Let us use the methods used before, i.e., let's find the semiperimeter of the triangle $A_1H_2F_1$

$$p = (a\sqrt{3}/2 + f + s)/2,$$

Height lowered from the vertex H_2

$$H = \frac{4}{a\sqrt{3}}|\sqrt{p(p - a\sqrt{3}/2)(p - f)(p - s)}|.$$

The unknown angle is found from the expression

$$\mathrm{Sin}B = \frac{H}{f}.$$

In case 3

$$\lambda = \frac{B_1H_2}{B_1F_1} = \frac{B_1C_{24}}{B_1H_{141}} = \frac{1}{\mathrm{Cos}\alpha_1}.$$

Then

$$B_1H_2 = r = \frac{a}{2\mathrm{Cos}\alpha_1}, F_1H_2 = f = \left|\sqrt{r^2 - a^2/4}\right|.$$

Let us consider the triangle $A_1B_1C_2$, in which all the sides are known, and find the section A_1H_2. From the point A_1 perpendicular is lowered to the side B_1C_2, the perpendicular base is denoted with K.

Semiperimeter

$$p = (a + l_3 + h_3 + l_7 + h_7)/2,$$

$$H = \frac{2}{l_3 + h_3}|\sqrt{p(p - a)(p - l_3 - h_3)(p - l_7 - h_7)}|, B_1K = v = \left|\sqrt{a^2 - H^2}\right|,$$

$$A_1H_2 = s = \left|\sqrt{H^2 + (r - v)^2}\right|.$$

Now in the triangle $A_1F_1H_2$, lengths of all sides are known, using the semiperimeter and the height, the unknown angle is found.

Semiperimeter

$$p = (a\sqrt{3}/2 + f + s)/2, H = \frac{2}{a\sqrt{3}/2}|\sqrt{p(p - a\sqrt{3}/2)(p - f)(p - s)}|,$$

The unknown linear angle equal to the dihedral angle $\angle B$ is found from the equation

$$\text{Sin}B = \frac{H}{f}$$

In case 4, the height is

$$F_1H_2 = F_1C_2 = f = \left|\sqrt{(l_6 + h_6)^2 - a^2/4}\right|.$$

As can be seen from the configuration, in this case, the dihedral angle is

$$\angle B = \pi/2. \tag{6.22}$$

In case 5—$\lambda = \frac{C_1H_2}{C_1F_1} = \frac{C_{23}C_1}{H_{13}C_1} = \frac{l_6+h_6}{a+|(l_3+h_3)Cos\alpha_1|}$,
Hence, $C_1H_2 = r = \frac{a/2(l_6+h_6)}{a+|(l_3+h_3)Cos\alpha_1|}$,
Then $F_1H_2 = f = \left|\sqrt{r^2 - a^2/4}\right|$.
Let us consider the triangle $A_1C_2C_1$, sides of which are known, and find the length of the section A_1H_2. For this purpose, the perpendicular is lowered from the vertex A_2 to the side C_2C_1, the intersection point is denoted with K

$$p = (a + l_6 + h_6 + l_7 + h_7)/2, A_1K = H = \frac{2}{l_6 + h}|\sqrt{p(p-a)(p-l_6-h)(p-l_7-h_7)}|,$$

$$C_1K = v = |\sqrt{a^2 - H^2}|, A_1H_2 = s = \left|\sqrt{H^2 + (r-v)^2}\right|.$$

Now, using the triangle $A_1H_2F_1$, in which all the sides are known, the unknown angle is found.

$$p = (a\sqrt{3}/2 + f + s_7)/2,$$
$$H = \frac{4}{\sqrt{3}}|\sqrt{p(p - a\sqrt{3}/2)(p-f)(p-s)}|, SinB = \frac{H}{f}. \tag{6.23}$$

In order to establish the coordinates of the point C_2 by transformation along the chain C_1–A_1–C_2, the link 7 is related to the moving coordinate system $A_1X_7Y_7Z_7$. The axis A_1Z_7 is directed along the edge A_1C_2, the axis A_1X_7—perpendicular to the plane of the face $A_1C_1C_2$ (Fig. 6.8). Let us perform the following sequence of motions to align the reference coordinate system with the system $A_1X_7Y_7Z_7$: parallel transfer of the system $C_1X_1Y_1Z_1$ along the axis $C_1X_1$for the value of $a_7 = a$, the new position of the reference system is denoted with A_1X_1YZ; rotation is performed for the angle of θ_7 ρ elative to the axis A_1Z until the axis A_1X_1 takes the position of A_1X, perpendicular to A_1Z_7; rotation is performed around the axis A_1X for the angle of α_7, until the coordinate axis A_1Z and A_1Z_7 coincide; rotation is performed for the angle of β_7 around the axis A_1Z_7 till the coincidence of the axis A_1X and A_1X_7, at the same time the coordinate systems fully coincide. The expression is obtained for the transformation parameters. The angle θ_7 is angle that is adjacent to one of the angles of the triangular angle at the vertex A_1 and it is found as shown above, according to

Fig. 6.8 Transformation parameters of the link 7

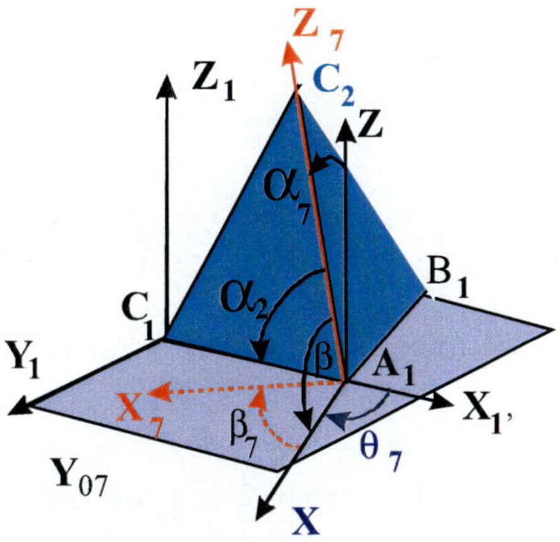

the cosine theorem for triangular angle.

$$\mathrm{Cos}\beta = \mathrm{Cos}\alpha_2 \cdot \mathrm{Cos}(\pi - \theta_7) + \mathrm{Sin}\alpha_2 \cdot \mathrm{Sin}(\pi - \theta_7) \cdot \mathrm{Cos}(\pi - A). \quad (6.24)$$

Given that $\beta = \pi/2$, after simplification, the following is obtained

$$Ctg\theta_7 = -tg\alpha_2\,\mathrm{Cos}A. \quad (6.25)$$

The angle α_2 at the base of the triangle $A_1C_1C_2$ is found as stated above by the formula $p = (a + l_7 + h_7 + l_6 + h_6)/2$; the height lowered from the vertex C_2—according to the equation $H = \frac{2}{a}|\sqrt{p(p-a)(p-l_7-h_7)(p-l_6-h_6)}|$ Then the angle α_2 equals

$$\alpha_2 = \arcsin(H/(l_7 + h_7)). \quad (6.26)$$

The dihedral angle A equals to the dihedral angle C that is defined by the Eq. (6.12). For the pyramid in question, transformation parameters table is made (Table 6.1). Hereinafter, expression for the function of transformation is used; it is obtained using *Mathcad* based on the transformation according to 8 parameters. After substituting appropriate transformation parameters from the Table 6.1 into the transformation function, transformation matrix is obtained. Matrix $\mathbf{A}_6^0$ is defined by the expression

$$\mathrm{F3}(0, -\theta6, 0, 0, -\alpha6, 0, 0, 0) \rightarrow \begin{pmatrix} \cos(\theta6) & \cos(\alpha6)\cdot\sin(\theta6) & \sin(\alpha6)\cdot\sin(\theta6) & 0 \\ -\sin(\theta6) & \cos(\alpha6)\cdot\cos(\theta6) & \cos(\theta6)\cdot\sin(\alpha6) & 0 \\ 0 & -\sin(\alpha6) & \cos(\alpha6) & 0 \\ 0 & 0 & 0 & 1 \end{pmatrix}$$

Table 6.1 Transformation parameters for the reference pyramid

Loops	Denotion transformation matrix $\mathbf{A}_i^0$	Transformation parameters							
		d_i	θ_i	γ_i	a_i	α_i	ψ_i	b_i	β_i
C_I–C_2	$\mathbf{A}_6^0$	0	$-\theta_6$	0	0	$-\alpha_6$	0	0	0
C–B_I	A_{03}^0	0	$\pi/3$	0	$a_3 = a$	0	0	0	0
B_I–C_2	$\mathbf{A}_3^{03}$	0	θ_3	0	0	$-\alpha_3$	0	0	0
C_I–A_I	A_{07}^0	0	0	0	a	0	0	0	0
A_I–C_2	$\mathbf{A}_7^{07}$	0	$-\theta_7$	0	0	α_7	0	0	0

Matrix A_{03}^0 is defined by the expression

$$F3\left(0, \frac{\pi}{3}, 0, a, 0, 0, 0, 0\right) \rightarrow \begin{pmatrix} \frac{1}{2} & -\frac{\sqrt{3}}{2} & 0 & \frac{a}{2} \\ \frac{\sqrt{3}}{2} & \frac{1}{2} & 0 & \frac{\sqrt{3}\cdot a}{2} \\ 0 & 0 & 1 & 0 \\ 0 & 0 & 0 & 1 \end{pmatrix}$$

For the chain B_I–C_2, elements of the matrix A_3^{03} equal to

$$F3(0, \theta 3, 0, 0, -\alpha 3, 0, 0, 0) \rightarrow \begin{pmatrix} \cos(\theta 3) & -\cos(\alpha 3)\cdot\sin(\theta 3) & -\sin(\alpha 3)\cdot\sin(\theta 3) & 0 \\ \sin(\theta 3) & \cos(\alpha 3)\cdot\cos(\theta 3) & \cos(\theta 3)\cdot\sin(\alpha 3) & 0 \\ 0 & -\sin(\alpha 3) & \cos(\alpha 3) & 0 \\ 0 & 0 & 0 & 1 \end{pmatrix}$$

For the chain C_I–A_I, elements of the matrix A_{07}^0 equal to:

$$F3(0, 0, 0, a, 0, 0, 0, 0) \rightarrow \begin{pmatrix} 1 & 0 & 0 & a \\ 0 & 1 & 0 & 0 \\ 0 & 0 & 1 & 0 \\ 0 & 0 & 0 & 1 \end{pmatrix}$$

For the chain A_I–C_2 matrix A_{07}^0 has the following form

$$F3(0, -\theta 7, 0, 0, \alpha 7, 0, 0, 0) \rightarrow \begin{pmatrix} \cos(\theta 7) & \cos(\alpha 7)\cdot\sin(\theta 7) & -\sin(\alpha 7)\cdot\sin(\theta 7) & 0 \\ -\sin(\theta 7) & \cos(\alpha 7)\cdot\cos(\theta 7) & -\cos(\theta 7)\cdot\sin(\alpha 7) & 0 \\ 0 & \sin(\alpha 7) & \cos(\alpha 7) & 0 \\ 0 & 0 & 0 & 1 \end{pmatrix}$$

Position of the connecting link 6 and coordinates of the node C_2 in the coordinate system $C_I X_I Y_I Z_I$ are found using the following matrix equations:

$$r_{C2}^0 = \mathbf{A}_6^0 r_6, r_{C2}^0 = (x_{C2} y_{C2} z_{C2} 1)^T, r_6 = (00 l_6 + h_6 1)^T, r_{C2}^0 = A_{03}^0 \mathbf{A}_3^{03} r_3,$$
$$r_3 = (00 l_3 + h_3 1)^T, r_{C2}^0 = A_{07}^0 \mathbf{A}_7^{07} r_7, r_Z = (00 l_Z + h_Z 1)^T. \quad (6.27)$$

The matrix Eq. (6.27) reduced to the following system of equations:

$$\begin{aligned}
&a)\, x_C = A \cdot \mathrm{Sin}(\alpha_6),\\
&b)\, y_C = B \cdot \mathrm{Sin}(\alpha_6),\\
&c)\, z_C = C \cdot \mathrm{Cos}(\alpha_6),\\
&d)\, x_C = \frac{1}{2}(a - D \cdot \mathrm{Sin}(\alpha_3),\\
&e)\, y_C = \frac{1}{2}(a\sqrt{3} + E \cdot \mathrm{Sin}(\alpha_3),\\
&f)\, z_C = (h_3 + l_3) \cdot \mathrm{Cos}(\alpha_3),\\
&g)\, x_C = a - F \cdot \mathrm{Sin}(\alpha_7),\\
&h)\, y_C = -G \cdot \mathrm{Sin}(\alpha_7),\\
&i)\, z_C = (h_7 + l_7) \cdot \mathrm{Cos}(\alpha_7).
\end{aligned} \tag{6.28}$$

where, *A*, *B*, *C*, *D*, *E*, *F*, *G* are equalities depending on known values, which look as follows

$$\begin{aligned}
&A = (h_6 + l_6) \cdot \mathrm{Sin}(\theta_6),\ B = (h_6 + l_6) \cdot \mathrm{Cos}(\theta_6),\ C = h_6 + l_6,\ D = (\mathrm{Sin}(\theta_3)\\
&\quad + \sqrt{3}\,\mathrm{Cos}(\theta_3))(h_3 + l_3),\ E = (\mathrm{Cos}(\theta_3 - \sqrt{3}\,\mathrm{Sin}(\theta_3)))(h_3 + l_3),\ F = (h_7 + l_7) \cdot \mathrm{Sin}(\theta_7),\\
&\quad G = (h_7 + l_7) \cdot \mathrm{Cos}(\theta_7).
\end{aligned} \tag{6.29}$$

In solving the direct kinematics problem, the known values are movements h_3, h_6, h_7 performed by the drives. The node C_2 has three degrees of freedom, therefore, its coordinates are uniquely defined by movements h_3, h_6, h_7. The solution of the systems of 9 transcendental equations define the angles α_3, α_6, α_7 between the axes of the connecting links and the axis C_1Z_1. Because the manipulator has redundant degrees of freedom representing the rotation angles around the axes of the connecting links, these angles can take arbitrary values and are unknown. The manipulator configuration does not depend on the rotation angles around the axes of the connecting links, i.e., the angles β_i. In this context, to simplify the solution, it is always assumed that the angle $\beta_i = 0$.

From the system of Eqs. (6.29), the unknown values are found:

$$\begin{aligned}
&\alpha_7 = arc\,\mathrm{Sin}\left(\frac{B \cdot a}{F \cdot B - G \cdot A}\right),\ x_C = a - F \cdot \frac{B \cdot a}{F \cdot B - G \cdot A},\\
&y_C = -G \cdot \frac{B \cdot a}{F \cdot B - G \cdot A},\ z_C = (h_7 + l_7) \cdot \mathrm{Cos}(\alpha_7),\ \alpha_6 = arc\,\mathrm{Sin}\left(\frac{x_C}{A}\right),\\
&\alpha_3 = arc\,\mathrm{Cos}\left(\frac{z_C}{h_3 + l_3}\right).
\end{aligned} \tag{6.30}$$

Position of the angle B_2 is determined by transforming the coordinate systems through two kinematic chains: via the paths C_1–B_2, C_1–B_1–B_2 and taking into account the constraint equation, which is obtained from the condition that the point B_2 is on the surface of the sphere with the radius C_2B_2. To determine the position of the point B_2, first the transformation (Fig. 6.9) is done along the chain C_1–B_2:

1. Rotation around the axis C_1Z_1 for the angle of θ_5 until the axis C_1X takes the position perpendicular to the axis C_1Z_5.
2. Rotation around the axis C_1X for the angle of α_5 until the axes C_1Z_1 and C_1Z_5 *coincide.*
3. Rotation for the angle of β_5 around the axis C_1Z_5 until the axis C_1X coincides with the axis C_1X_5.

Let consider the trihedral angle with vertex at the point C_1 and faces C_1a, C_1b, C_1c. The face angle aC_1c is denoted with β. Other face angles are denoted in Fig. 6.9.

According to the theorem of cosines, for the trihedral angle with the vertex C_1 there is

$$\mathrm{Cos}\,\beta = \mathrm{Cos}\,\alpha_8 \cdot \mathrm{Cos}\,(\pi/3 + \theta_5) + \mathrm{Sin}\,\alpha_8 \cdot \mathrm{Sin}\,(\pi/3 + \theta_{53}) \cdot \mathrm{Cos}\,C_1.$$

where $\beta = \pi/2$, is dihedral angle $\angle C_1$ formed by the surface $C_1B_2B_1$ with the plane $C_1X_1Y_1$ of the base. After simplification, the following is obtained

$$Ctg(\pi/3 + \theta_5) = -tg\alpha_8 CosC_1. \tag{6.31}$$

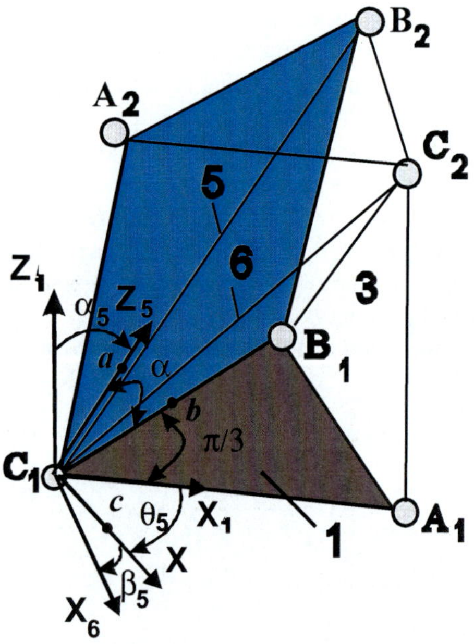

Fig. 6.9 Transformation of the coordinate systems along the chain C_1–B_2

To determine the angle α_8 at the base of the face $C_1B_2B_1$, semiperimeter of the triangle $C_1B_2B_1$ *is found*

$$p = (a + h_5 + l_5 + h_8 + l_8)/2. \tag{6.32}$$

Let us define the length of the height lowered from the vertex B_2

$$H = \frac{2}{a}|\sqrt{p(p-a)(p-h_5-l_5)(p-h_8-l_8|}. \tag{6.33}$$

The angle α_8 at the base of the triangle equals to

$$\alpha_8 = arc\,\mathrm{Sin}(H/(l_5 + h_5). \tag{6.34}$$

When determining the dihedral angle $\angle C_1$, linear angle between the lines F_1F_2 and F_1A_1 that are perpendicular to the side C_1B_1 is considered. Due to changes in the manipulator configuration, there are three possible cases of positioning of the point F_2 (Fig. 6.10).

In the 1st case of $(l_5 + h_5)Cos\alpha_8 > a/2$, the straight line F_1F_2 crosses the edge C_1B_2 (Fig. 6.10*a*). In the 2nd case of $(l_5 + h_5)Cos\alpha_8 = a/2$, the point F_2 coincides with the point B_2 (Fig. 6.10*b*). In the 3rd case of $(l_5 + h_5)\,\mathrm{Cos}\,\alpha_8 < a/2$, the straight line F_1F_2 crosses the edge B_1B_2 (Fig. 6.10c).

Each of these cases uses its own algorithm for determining the dihedral angle $\angle C_1$.*1 case*. From the point B_2, the perpendicular C_2G_1 is lowered to the side C_1B_2.

Let us make-up a ratio

$$\lambda = \frac{C_1F_2}{C_1F_1} = \frac{C_1B_2}{C_1G_1} = \frac{1}{\mathrm{Cos}\,\alpha_8}, C_1B_2,$$

Let us determine

$$C_1F_2 = r = \frac{a/2}{\mathrm{Cos}\,\alpha_8}, \; F_1F_2 = f = \frac{a}{2}tg\alpha_8.$$

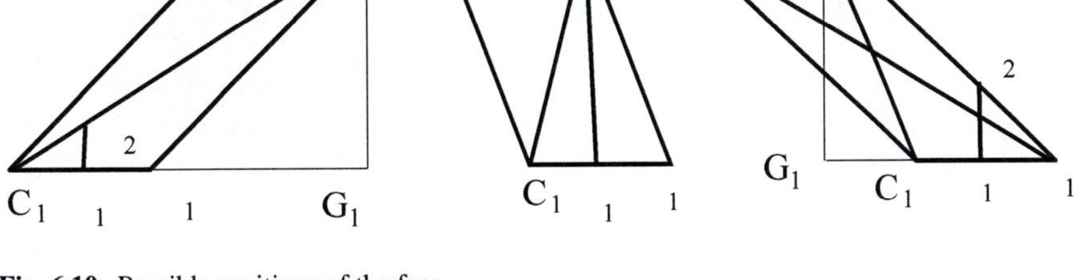

Fig. 6.10 Possible positions of the face

Now, from the expression (6.31), it follows

$$\theta_3 = arcctg(-tg\alpha_8 \operatorname{Cos} C_1) - \pi/3.$$

To transform the coordinate system along the chain C_1–B_1–B_2, it is needed to select

the coordinate system $B_1X_8Z_8$ related to the link 8 (Fig. 6.11) so that the axis B_1Z_8 is directed along the link 8, and the axis B_1X_8 is located in the plane of the face $A_1B_1B_2C_2$. Then the following sequence of actions is performed to align the reference coordinate system with the moving system related to the 8 link:

1. Rotation of the system $C_1X_1Z_1$ around the axis C_1Z_1 for the angle of $\theta = \pi/3$, until the axis direction coincides with the side C_1B_1.
2. Transfer is performed along C_1B_1 for the distance of a. The new position of the coordinate system is $B_1X\,Z$.
3. Rotation around the axis B_1Z for the angle of θ_8, until the axis B_1X takes the position of X_{08}, that is perpendicular to the axis B_1Z_8;
4. Rotation around the axis B_1X_{08} for the angle of α_8 till the coincidence of the axis B_1Z and B_1Z_8;
5. Rotation for the angle of β_8 around the axis B_1Z_8, until the axis B_1X_{08} coincides with the axis B_1X_8.

Similarly to the above, the angle θ_8 is found from the expression for the cosine of the triangular angle formed by the planes with face angles $\alpha_{10}\beta_{10}\gamma_{10}$ (Fig. 6.11).

According to the theorem of cosines for this trihedral angle with the vertex B_1, the following is obtained

$$\operatorname{Cos}\beta_{10} = \operatorname{Cos}\alpha_{10} \cdot \operatorname{Cos}\gamma_{10} + \operatorname{Sin}\alpha_{10} \cdot \operatorname{Sin}\gamma_{10} \cdot \operatorname{Cos}B_1. \qquad (6.35)$$

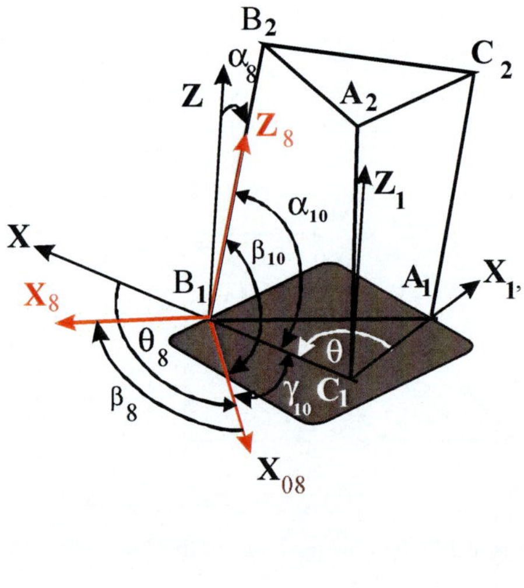

Fig. 6.11 Transformation along the Chain C_1–B_1–B_2

where, $\beta_{10} = \pi/2$, angle $\gamma_{10} = \pi - \theta_8$, dihedral angle $\angle B_1$ formed by the surface $C_1B_2B_1$ and the plane $C_1X_1Y_1$ of the base equals $\angle C_1$ and is found above.

From the expression (6.35) the following is obtained

$$Ctg(\theta_8) = tg\alpha_8 \, \mathrm{Cos} B_1.$$

To determine the angle α_8 at the base of the face $C_1B_2B_1$, semiperimeter of the triangle $C_1B_2B_1$ *is found*

$$p = (a + h_5 + l_5 + h_8 + l_8)/2.$$

Let us define the length of the height lowered from the vertex B_2

$$H = \frac{2}{a} |\sqrt{p(p-a)(p-h_5-l_5)(p-h_8-l_8}|.$$

The angle α_8 at the base of the triangle equals

$$\alpha_8 = \arcsin(H/(l_5 + h_5).$$

The resulting transformation parameters are summarized in Table 6.2.

Position of the angle B_2 is determined by joint solution of the vector equations

$$\begin{aligned} r^0_{B2} &= A^0_5 r_5, \boldsymbol{r}_5 = (00 l_5 + h_s 1)^T, \\ r^0_{B2} &= A^0_{08} \mathbf{A}^{08}_8 r_8, r_8 = (00 l_8 + h_8 1)^T. \end{aligned} \tag{6.36}$$

Since the point B_2 is located on the surface of a sphere with a radius a with the center at the point C_2, then additionally to the Eq. (6.36), the sphere equation is used

$$(x_{C2} - x_{B2})^2 + (y_{C2} - y_{B2})^2 + (z_{C2} - z_{B2})^2 - a^2 = 0.$$

The matrix Eq. (6.36) can be represented as

Table 6.2 Transformation parameters for determining the position of the node B_2

Loops	Notation of the transformation matrix $\mathbf{A}^0_i$	Transformation parameters							
		d_i	θ_i	γ_i	a_i	α_i	ψ_i	b_i	β_i
C_1–B_2	$\mathbf{A}^0_5$	0	θ_5	0	0	α_5	0	0	β_5
C_1–B_1	A^0_{08}	0	$\pi/3$	0	a	0	0	0	0
B_1–B_2	A^{08}_8	0	θ_8	0	0	α_8	0	0	β_8

$$\begin{pmatrix} \text{xB2} \\ \text{yB2} \\ \text{zB2} \\ 1 \end{pmatrix} = \begin{bmatrix} \sin(\alpha 5)\cdot\sin(\theta 5)\cdot(h5+15) \\ -\cos(\theta 5)\cdot\sin(\alpha 5)\cdot(h5+15) \\ \cos(\alpha 5)\cdot(h5+15) \\ 1 \end{bmatrix}$$

$$\begin{pmatrix} \text{xB2} \\ \text{yB2} \\ \text{zB2} \\ 1 \end{pmatrix} = \begin{bmatrix} \frac{a}{2}+\left(\frac{\sin(\alpha 8)\cdot\sin(\theta 8)}{2}+\frac{\sqrt{3}\cdot\cos(\theta 8)\cdot\sin(\alpha 8)}{2}\right)\cdot(\text{h8}+18) \\ \frac{\sqrt{3}\cdot a}{2}-\left(\frac{\cos(\theta 8)\cdot\sin(\alpha 8)}{2}-\frac{\sqrt{3}\cdot\sin(\alpha 8)\cdot\sin(\theta 8)}{2}\right)\cdot(\text{h8}+18) \\ \cos(\alpha 8)\cdot(\text{h8}+18) \\ 1 \end{bmatrix} \tag{6.37}$$

Equations (6.37) can be represented as

$$\begin{aligned} x_{B2} &= A1\mathrm{Sin}(\alpha_5), \\ y_{B2} &= B1\mathrm{Sin}(\alpha_5), \\ z_{B2} &= (h_5+l_5)\mathrm{Cos}(\alpha_5), \\ x_{B2} &= a/2+\frac{\mathrm{Sin}(\alpha_8)}{2}C1, \\ y_{B2} &= \sqrt{3}a/2-\frac{\mathrm{Sin}(\alpha_8)}{2}D1, \\ z_{B2} &= (h_8+l_8)\mathrm{Cos}(\alpha_8). \end{aligned} \tag{6.38}$$

The following notation for the expressions are applied here:

$$\begin{aligned} A1 &= \mathrm{Sin}(\theta_5)\cdot(h_5+l_5), \\ B1 &= -\mathrm{Cos}(\theta_5)\cdot(h_5+l_5), \\ C1 &= [\mathrm{Sin}(\theta_8+\sqrt{3}\mathrm{Cos}(\theta_8)](h_8+l_8), \\ D1 &= [\mathrm{Cos}(\theta_8)-\sqrt{3}\mathrm{Sin}(\theta_8](h_8+l_8). \end{aligned} \tag{6.39}$$

As a result of solving systems of Eqs. (6.38) taking into account (6.39), the following is obtained

$$\alpha_8 = arc\mathrm{Sin}\left[\frac{a(A1\sqrt{3}-B1}{A1\cdot D1+B1\cdot C1}\right], x_{B2} = a/2+\frac{C1}{2}\left[\frac{a(A1\sqrt{3}-B1}{A1\cdot D1+B1\cdot C1}\right],$$

$$y_{B2} = \frac{a\sqrt{3}}{2}-\frac{D1}{2}\left[\frac{a(A1\sqrt{3}-B1}{A1\cdot D1+B1\cdot C1}\right], z_{B2} = (h_8+l_8)\left|\sqrt{1-\left[\frac{a(A1\sqrt{3}-B1}{A1\cdot D1+B1\cdot C1}\right]^2}\right|,$$

$$\alpha_5 = arc\mathrm{Sin}\left(\frac{x_{B2}}{A1}\right). \tag{6.40}$$

Equation (6.36) is used to verify the results of calculating the unknown values by the dependencies (6.40). To determine the position of the node A_2 in the coordinate system $C_1X_1Y_1Z_1$, the coordinate system (Fig. 6.12) is transformed along the chain

Fig. 6.12 Transformation parameters of the link 4

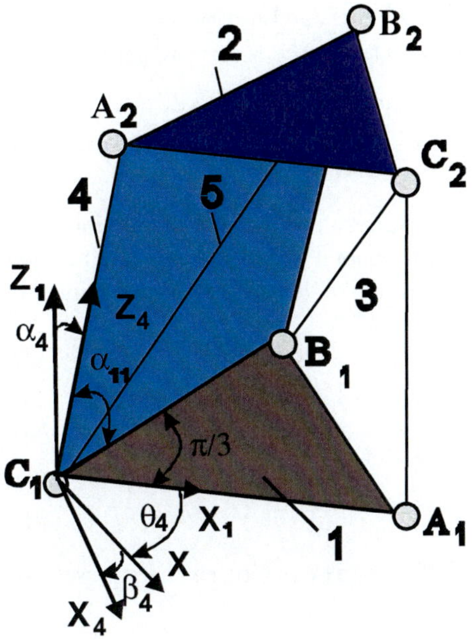

C_1–A_2. The coordinate system related to the moving links, are selected similarly to the above examples. According to the theorem of cosines of the trihedral angle with faces C_1A_2, C_1B_1, C_1X, the following is obtained

$$\mathrm{Cos}\,\beta_{11} = \mathrm{Cos}\,\alpha_{11} \cdot \mathrm{Cos}\gamma_{11} + \mathrm{Sin}\,\alpha_{11} \cdot \mathrm{Sin}\gamma_{11} \cdot \mathrm{Cos}C_1 \tag{6.41}$$

where $\beta_{11} = \pi/2$, the angle $\gamma_{11} = \theta_4 + \pi/3$, dihedral angle $\angle C_1$ formed by the surface $C_1A_2\, B_2B_1$ with the plane $C_1B_1A_1$ of the base, is found above. After simplifying the expression (6.41), the following is obtained

$$Ctg\gamma_{11} = -tg\alpha_{11}\mathrm{Cos}C_1, \theta_4 = arctg(-tg\alpha_{11}\mathrm{Cos}C_1) - \pi/3.$$

The angle α_{11} is found from the properties of the scalar product of the vectors C_1A_2 *and* C_1B_1 set by the coordinates, i.e.

$$\mathrm{Cos}\,\alpha_{11} = \frac{x_{A2}/2 + y_{A2}\sqrt{3}/2}{l_4 + h_4}$$

Then

$$tg\alpha_{11} = \sqrt{\frac{1}{\mathrm{Cos}^2\alpha_{11}} - 1}.$$

Table 6.3 Transformation parameters for determining the position of the node A_2

Loops	Designation of the transformation matrix $\mathbf{A}_i^0$	Transformation parameters							
		d_i	θ_i	γ_i	a_i	α_i	ψ_i	b_i	β_i
C_1–A_2	A_4^0	0	θ_4	0	0	α_4	0	0	β_4

As a result of the transformations that are similar to those made above, the transformation parameters are obtained and summarized in Table 6.3.

Position of the angle A_2 is determined by the matrix equation

$$r_{A2}^0 = A_4^0 r_4,\ r_{A2}^0 = (x_{A2} y_{A2} z_{A2} 1)^T,\ r_4 = (00 l_4 + h_4 1)^T,$$

$$\begin{pmatrix} xA2 \\ yA2 \\ zA2 \\ 1 \end{pmatrix} = \begin{bmatrix} \sin(\alpha 4) \cdot \sin(\theta 4) \cdot (h4 + 14) \\ -\cos(\theta 4) \cdot \sin(\alpha 4) \cdot (h4 + 14) \\ \cos(\alpha 4) \cdot (h4 + 14) \\ 1 \end{bmatrix}$$

The matrix equation can be represented as a system of equations

$$\begin{aligned} x_{A2} &= L4 \cdot \mathrm{Sin}(\theta_4)\mathrm{Sin}(\alpha_4), \\ y_{A2} &= L4 \cdot \mathrm{Cos}(\theta_4)\mathrm{Sin}(\alpha_4), \\ z_{A2} &= L4 \cdot \mathrm{Cos}(\alpha_4), \end{aligned} \tag{6.42}$$

where $L4 = l_4 + h_4$.

To determine the position of the point A_2, two additional constraint equations are used; these equations are the result of that the point A_2 is at the intersection of two spherical surfaces represented by the equations

$$\begin{aligned} &(x_{C2} - x_{A2})^2 + (y_{C2} - y_{A2})^2 + (z_{C2} - z_{A2})^2 - a^2 = 0, \\ &(x_{A2} - x_{B2})^2 + (y_{A2} - y_{B2})^2 + (z_{A2} - z_{B2})^2 - a^2 = 0 \end{aligned} \tag{6.43}$$

The system of Eq. (6.43) can be simplified if the second equation is subtracted from the 1st one. Then, after a series of reforms, the following is obtained

$$A2 \cdot x_{A2} + B2 \cdot y_{A2} + C2 \cdot z_{A2} - D2 = 0,$$

where

$$A2 = x_{C2} - x_{B2},\ B2 = y_{C2} - y_{B2},\ C2 = z_{C2} - z_{B2},\ D2 = \frac{(l_6 + h_6)^2 - (l_5 + h_5)^2}{2}.$$

Having found the coordinates y_{A2}, z_{A2} from the Eq. (6.42) via x_{A2}, from the Eq. (6.43) the following is obtained

$$A2 \cdot x_{A2} + B2 \cdot \frac{x_{A2}}{tg(\theta_4)} + C2 \cdot \frac{\sqrt{(L4 \cdot \mathrm{Sin}(\theta_4))^2 - x_{A2}^2}}{\mathrm{Sin}(\theta_4)} - D2 = 0.$$

$$E2 = \frac{D2 \cdot \mathrm{Sin}\theta_4}{C2}, \, G2 = \left(A2 + \frac{B2}{tg\theta_4}\right)\frac{\mathrm{Sin}\theta_4}{C2}.$$

Then

$$(E2 - G2 \cdot x_{A2})^2 = (L4 \cdot \mathrm{Sin}\theta_4)^2 - x_{A2}^2.$$

After introducing the additional symbols in the form

$$F2 = (G2)^2 + 1, \; H2 = 2 \cdot E2 \cdot G2, \; V = (L4 \cdot \mathrm{Sin}\,\theta_4)^2 - (E2)^2$$

the quadratic equation of the following form is obtained

$$(F2) \cdot x_{A2}^2 - H2 \cdot x_{A2} - V = 0,$$

solution for which has the following form

$$(x_{A2})_{1,2} = \frac{-H2 \pm \sqrt{(H2)^2 - 4 \cdot F2 \cdot V}}{2 \cdot F2}.$$

Of the two roots of the equation, the closest to the previous value of the coordinates x_{A2} is taken. Other unknown values are determined from the system (6.42):

$$\mathrm{Sin}\alpha_4 = \frac{x_{A2}}{L4 \cdot \mathrm{Sin}\theta_4}, \, \alpha_4 = arc\mathrm{Sin}(\frac{x_{A2}}{L4 \cdot \mathrm{Sin}\theta_4}),$$
$$y_{A2} = x_{A2} \cdot Ctg\theta_4, \, z_{A2} = L4 \cdot \left|\sqrt{1 - (\mathrm{Sin}\alpha_4)^2}\right|.$$

Thus, position of the node C_2 is determined by 3 independent parameters h_3, h_6, h_7; position of the node B_2 is determined by 2 parameters h_5, h_8; position of the node A_2 is determined by one parameter h_4. Other transformation parameters depend on the manipulator configuration. While the angles α_3, α_6, α_7, α_5, α_8, α_4, are the unknown angles required for controlling the manipulator functioning as an adaptive support.

Based on the dependencies obtained above, an algorithm has been formed and a program has been compiled in Matlab; the program allows determining positions of the upper platform relative to the bottom one by the change in the motion of the rod of each connecting link. As an example, Fig. 6.13 shows the following: (*a*) diagram of the spherical displacement of the platform with the fixed point C_2; (b) spatial translational movement with operation of all the drives for the same values; (c) motions of the platform during different movements in drives; (d) motions performed only by diagonal drives; (e) simultaneous movements by sides for greater values than

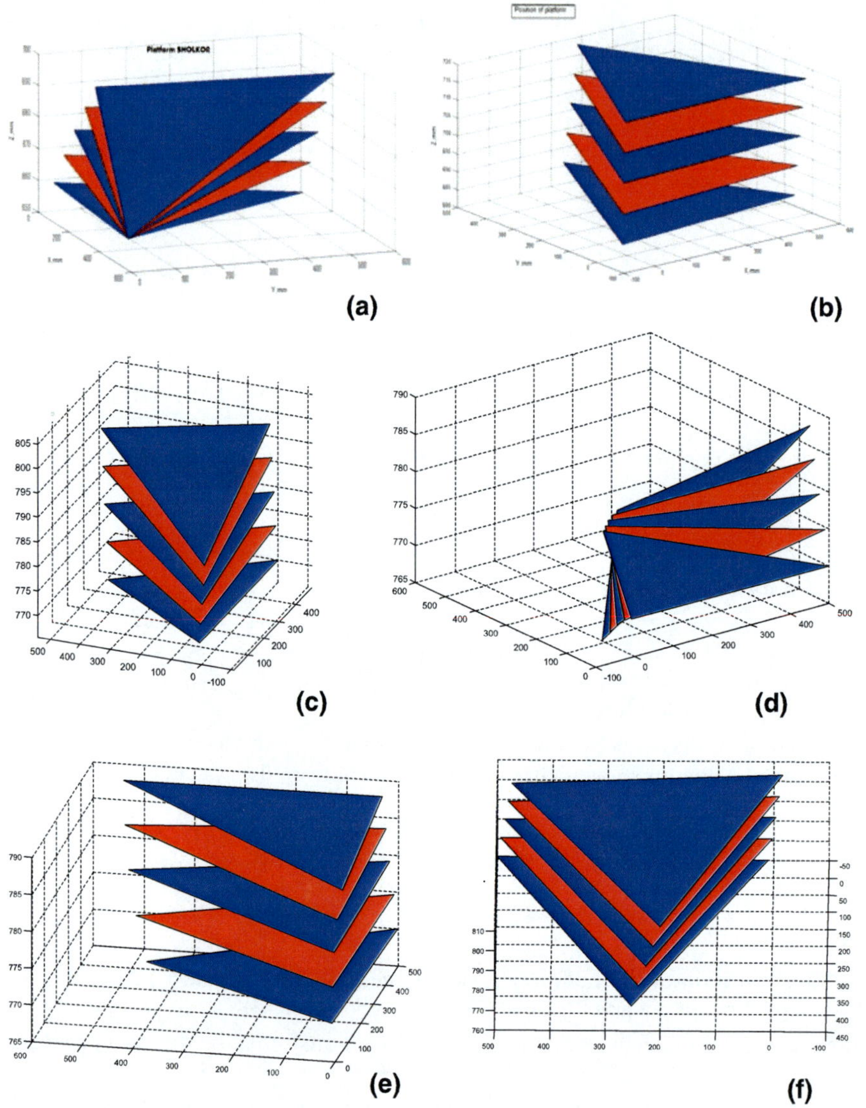

Fig. 6.13 Platform positions research

the diagonal ones; (f) different by diagonals for greater values than the sides. Using computer modeling, it is also found that movement performed by a certain driver can move the node point over the cylindrical surface, curvature of which depends on the lengths of the other two links connected in this node; minimum and maximum length of the connecting link can be selected individually reasoning from the requirements to the working medium; decrease of the lengths of the triangle sides results in the reduction of curvature of the cylindrical surface representing a multitude of node points provisions; by selecting the laws of changing the lengths of the connecting links, it is possible using the provided program to obtain any predetermined motion of the platform.

6.3 Solution a Inverse Position Problem of the Platform for the Multiloop Parallel Manipulator

The inverse task of Position Problem of the Platform for the Multiloop Manipulator is to determine the movement of the actuators and their orientation in space at the specified positions of the mobile platform. The following results were obtained in [49] in relation to a sailing wind power station (WPS) using the MPM «SHOLKOR». To describe the algorithm, the computational model (Fig. 6.14). Figure 6.14 presents the general design model of the WPS consisting of bottom 1 and top 2 platforms of the manipulator connected by the actuators 4–8 by flexible cable bonds: 9—made of two links; 10—made of three links; 11—made of four links. The mast 12 rigidly connects the sail 13 with the upper platform 2 of the manipulator. To analyze the working body (WB) dynamics, two coordinate systems have been accepted. The fixed coordinate system $C_1X_0Y_0Z_0$ is related to the fixed platform 1 of the manipulator, the axis C_1Z_0 is directed vertically upward, and the axis C_1Y_0 is directed along the edge C_1A_1.

The start of the movable system $O_{01}X_{01}Y_{01}Z_{01}$ coincides with the center of the movable platform in the initial position O_{01}. The axis $O_{01}Z_1$ is parallel to the axis C_1Z_0. First the task of determining the position of nodes A_2, B_2, C_2 of the top platform is set, if the position of the WB is set by the coordinates ($\Delta y_S, \Delta z_S$) of the center S and rotation ($\Delta\varphi_1$) relative to the axis crossing S perpendicularly to the Q *plane.* Initial data are considered to be that the upper and bottom platform in present regular triangles with sides equal to a. Platform height in its initial position amounts to H. To solve this kinematic problem, we use homogeneous transformation matrix applied in robotics [47]. As noted above: the reference system $C_1X_0Y_0Z_0$ is fixed; auxiliary fixed system $O_{01}X_{01}Y_{01}Z_{01}$ with its beginning in the center of the top platform has axes that are parallel to the axes of the system $C_1X_0Y_0Z_0$. Also, the $O_{01}X_1Y_1Z_1$ system is introduced, it is related to the Q plane depending on the wind direction. The current WB position is presented by the movable triangular prism SA_2, B_2, C_2. Let us define the coordinates of the top platform nodes in the fixed system $O_{01}X_{01}Y_{01}Z_{01}$. Based on the condition of selecting the auxiliary fixed coordinate system, it follows that in the platform initial position, coordinates A_2, B_2, C_2 are determined by the radius vectors

$$r_{A2} = \begin{pmatrix} 0 & 0 & H & 1 \end{pmatrix}^T, r_{B2} = \begin{pmatrix} a\frac{\sqrt{3}}{2} & a/2 & H & 1 \end{pmatrix}, r_{C2} = \begin{pmatrix} 0 & a & H & 1 \end{pmatrix}^T. \quad (6.43)$$

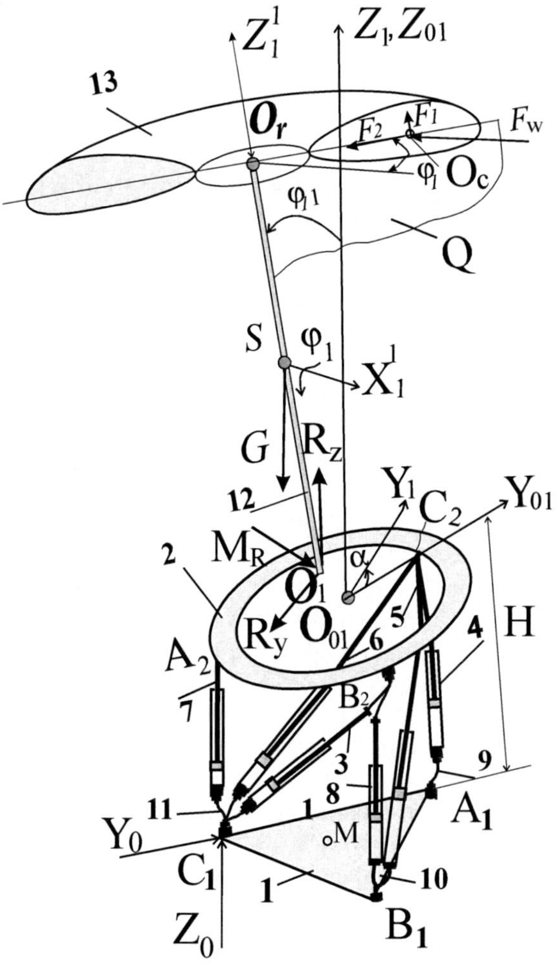

Fig. 6.14 The computational model of the sail WPS

Coordinates of the nodes in the local coordinate system $O_{01}X_1Y_1Z_1$ obtained transfer for the value of H along the axis $O_{01}Z_{01}$, for the value of $a/2$ along the axis $O_{01}Y_{01}$ and a shift for the value of $a\sqrt{3}/6$ along the axis $O_{01}X_{01}$, by rotating the system around the axis $O_{01}Z_{01}$ for the angle of α are determined according to the equation

$$r^1_{A2} = A^{01}_1 r_{A2},\ r^1_{B2} = A^{01}_1 r_{B2},\ r^1_{C2} = A^{01}_1 r_{C2}, \tag{6.44}$$

where transformation matrix is

$$A_1^{01} = \begin{pmatrix} Cos\alpha & Sin\alpha & 0 & -a/2 \\ -Sin\alpha & Cos\alpha & 0 & -a\sqrt{3}/6 \\ 0 & 0 & 1 & -H \\ 0 & 0 & 0 & 1 \end{pmatrix}. \quad (6.45)$$

Transformation matrix in case of parallel transfer of the mobile system for the value of Δy_s, Δz_s and a shift for the value of L_S along the $O_I Z_I^I$ axis look as follows

$$A_T = \begin{pmatrix} 1\,0\,0 & 0 \\ 0\,1\,0 & \Delta y_S^i \\ 0\,0\,1 & \Delta z_S^i + L_S \\ 0\,0\,0 & 1 \end{pmatrix}. \quad (6.46)$$

Rotation around the SX_1^1 axis for the angle of $\Delta\varphi_1^i$ is described by the matrix

$$A_R = \begin{pmatrix} 1 & 0 & 0 & 0 \\ 0 & Cos\Delta\varphi_1^i & -Sin\Delta\varphi_1^i & 0 \\ 0 & Sin\Delta\varphi_1^i & Cos\Delta\varphi_1^i & 0 \\ 0 & 0 & 0 & 1 \end{pmatrix}. \quad (6.47)$$

Matrix of parallel transfer along axis $S_1 Z_1^1$ for the value of -L_S

$$A_L = \begin{pmatrix} 1\,0\,0 & 0 \\ 0\,1\,0 & 0 \\ 0\,0\,1 & -L_S \\ 0\,0\,0 & 1 \end{pmatrix}. \quad (6.48)$$

To determine positions of the nodes of the top platform and its center O_1 in the coordinate system $O_{0I}X_I Y_I Z_I$ with slight displacements of the WB, the following transformation matrix is used (6.46–6.48)

$$T_1^{11} = A_T \cdot A_R \cdot A_L. \quad (6.49)$$

Now, the position of the nodes and the center O_1 of the platform is the local system $O_{0I}X_I Y_I Z_I$ is determined by the equation

$$r_{A2}^{01} = T_1^{11} r_{A2}^1,\; r_{B2}^{01} = T_1^{11} r_{B2}^1,\; r_{C2}^{01} = T_1^{11} r_{C2}^1,\; r_{O1}^{01} = T_1^{11} (0\,0\,0\,1)^T. \quad (6.50)$$

The coordinates of these points in the fixed system $C_I X_0 Y_0 Z_0$ amount to

$$r_{A2}^0 = T_0^1 r_{A2}^{01},\; r_{B2}^0 = T_0^1 r_{B2}^{01},\; r_{C2}^0 = T_0^1 r_{C2}^{01},\; r_{O1}^0 = T_0^1 r_{O1}^{01}, \quad (6.51)$$

where the transformation matrix is defined as inverse matrix (6.48)

$$T_0^1 = (A_1^{01})^{-1} \tag{6.52}$$

A change in the position of the upper platform of the manipulator causes a change in the position, lengths of actuators, the direction of reaction forces vectors effecting the upper platform from the actuators. To determine these values, the inverse kinematics problem is solved for the manipulator. This problem consists in that at a known position of the moving upper platform on coordinates of points of junctions (A_2, B_2, C_2), the orientation and lengths of actuators are established, hence the direction of actuators reaction forces acting along the axis of actuators. As an example, let us solve the inverse problem for the actuator 6. The coordinates of the junction C_2 ($x_{C2}^0\ y_{C2}^0\ z_{C2}^0$) are known from the Eq. (6.9), the coordinates of the junction C1 (0 0 0) are equal to zero. The current length of the actuator 6 is equal to

$$l_{6t} = \sqrt{(x_{C2}^0)^2 + (y_{C2}^0)^2 + (z_{C2}^0)^2}. \tag{6.53}$$

Angles $\alpha_6, \beta_6, \gamma_6$ between the axis of the actuator 6 and axes C_1X_0, C_1Y_0, C_1Z_0 of the fixed coordinate system are determined as

$$\alpha_6 = arc\cos(x_{C2}^0/l_{6t}),\ \beta_6 = arc\cos(y_{C2}^0/l_{6t}),\ \gamma_6 = arc\cos(z_{C2}^0/l_{6t}). \tag{6.54}$$

Change in the length of the actuator 6 is

$$\Delta l_6 = l_{6t} - l_6. \tag{6.55}$$

If angles $\alpha_6, \beta_6, \gamma_6$ are known, then it is easy to determine, as it is shown below, the projections of the reaction forces N_{6X}, N_{6Y}, N_{6Z} of the actuator 6 to the fixed coordinate axis. For other actuators, similar calculations are made.

The coordinates of the nodes are defined by the formula (6.52) as follows

$$r_{A2}^0 = \begin{pmatrix} x_{A2}^0 \\ y_{A2}^0 \\ z_{A2}^0 \\ 1 \end{pmatrix},\ r_{B2}^0 = \begin{pmatrix} x_{B2}^0 \\ y_{B2}^0 \\ z_{B2}^0 \\ 1 \end{pmatrix}\ r_{C2}^0 = \begin{pmatrix} x_{C2}^0 \\ y_{C2}^0 \\ z_{C2}^0 \\ 1 \end{pmatrix},\ r_{O1}^0 = \begin{pmatrix} x_{O1}^0 \\ y_{O1}^0 \\ z_{O1}^0 \\ 1 \end{pmatrix} \tag{6.56}$$

The current value of the actuators length is determined from the following expression

$$l_{3t} = \sqrt{(x_{B2}^0 - x_{C1})^2 + (y_{B2}^0 - y_{C1})^2 + (z_{B2}^0 - z_{C1})^2},$$

$$l_{4t} = \sqrt{(x_{C2}^0 - x_{A1})^2 + (y_{C2}^0 - y_{A1})^2 + (z_{C2}^0 - z_{A1})^2},$$

$$l_{5t} = \sqrt{(x_{C2}^0 - x_{B1})^2 + (y_{C2}^0 - y_{B1})^2 + (z_{C2}^0 - z_{B1})^2},$$

$$l_{6t} = \sqrt{(x_{C2}^0 - x_{C1})^2 + (y_{C2}^0 - y_{C1})^2 + (z_{C2}^0 - z_{C1})^2},$$

$$l_{7t} = \sqrt{(x_{B2}^0 - x_{C1})^2 + (y_{B2}^0 - y_{C1})^2 + (z_{B2}^0 - z_{C1})^2},$$
$$l_{8t} = \sqrt{(x_{B2}^0 - x_{B1})^2 + (y_{B2}^0 - y_{B1})^2 + (z_{B2}^0 - z_{B1})^2}. \tag{6.57}$$

Here, the coordinates of the bottom platform nodes are known, i.e.

$$x_{C1} = y_{C1} = z_{C1} = 0;\ x_{A1} = z_{A1} = 0,\ y_{A1} = a;\ x_{B1} = a \cdot \sqrt{3}/2,$$
$$y_{B1} = a/2,\ z_{B1} = 0 \tag{6.58}$$

It should be noted that in the initial position, the edges lengths (of the actuators) are equal to

$$l_{30} = l_{50} = l_{60} = A,$$
$$l_{40} = l_{70} = l_{80} = H, \tag{6.59}$$

where H is the manipulator height (set); $A = \sqrt{H^2 + a^2}$ is the length of the diagonal line in its initial position.

Change in the lengths of the edges is equal to

$$\Delta l_3 = l_{3t} - l_{30},\ \Delta l_4 = l_{4t} - l_{40},\ \Delta l_5 = l_{5t} - l_{50},\ \Delta l_6 = l_{6t} -,\ l_{60}$$
$$\Delta l_7 = l_{7t} - l_{70},\ \Delta l_8 = l_{8t} - l_{80}. \tag{6.60}$$

Angles (α,β,γ) between the fixed axes C_1X_0, C_1Y_0, C_1Z_0, respectively, and the axes of the six actuators (3–8) are determined from the equations

$$\alpha_3 = \arccos\Big((x_{B2}^0 - x_{C1})/l_{3t}\Big),\ \beta_3 = \arccos\Big((y_{B2}^0 - y_{C1})/l_{3t}\Big),\ \gamma_3 = \arccos\Big((z_{B2}^0 - z_{C1})/l_{3t}\Big);$$
$$\alpha_4 = \arccos\Big((x_{C2}^0 - x_{A1})/l_{4t}\Big),\ \beta_4 = \arccos\Big((y_{C2}^0 - y_{A1})/l_{4t}\Big),\ \gamma_4 = \arccos\Big((z_{C2}^0 - z_{A1})/l_{4t}\Big);$$
$$\alpha_5 = \arccos\Big((x_{C2}^0 - x_{B1})/l_{5t}\Big),\ \beta_5 = \arccos\Big((y_{C2}^0 - y_{B1})/l_{5t}\Big),\ \gamma_5 = \arccos\Big((z_{C2}^0 - z_{B1})/l_{5t}\Big);$$
$$\alpha_6 = \arccos\Big((x_{C2}^0 - x_{C1})/l_{6t}\Big),\ \beta_6 = \arccos\Big((y_{C2}^0 - y_{C1})/l_{6t}\Big),\ \gamma_6 = \arccos\Big((z_{C2}^0 - z_{C1})/l_{6t}\Big);$$
$$\alpha_7 = \arccos\Big((x_{A2}^0 - x_{C1})/l_{7t}\Big),\ \beta_7 = \arccos\Big((y_{A2}^0 - y_{C1})/l_{7t}\Big),\ \gamma_7 = \arccos\Big((z_{A2}^0 - z_{C1})/l_{7t}\Big);$$
$$\alpha_8 = \arccos\Big((x_{B2}^0 - x_{B1})/l_{8t}\Big),\ \beta_8 = \arccos\Big((y_{B2}^0 - y_{B1})/l_{8t}\Big),\ \gamma_8 = \arccos\Big((z_{B2}^0 - z_{D1})/l_{8t}\Big);$$

Based on the specified algorithm, program was designed and computer calculations were made. While, it was accepted that the initial data are as follows: capacity of the sail WPS $N = 5$ kWt; wind speed $v = 3$ m/s; mast length $L = 10$ m; dimensions of the side of the regular top and bottom triangles are $a = 1$ m. This calculations are made for the WB positions corresponding to the rotation for the angle of $\Delta\varphi_i = i \cdot \beta/N, i = 1, \ldots, N$, here $\beta = 1°$ is the current angle of the mast deviation from the vertical position; with $N = 10$ the maximal deviation angle of the WB equal to 9° will be reached. The maximum angular velocity of the WB depends on the

Table 6.4 The displacement in the actuator

degr.	Actuators, m					
	No. φ					
	3	4	5	6	7	8
0	0	0	0	0	0	0
1	5.548×10^{-3}	0.021	0.016	0.016	5.548×10^{-3}	7.353×10^{-3}
2	0.014	0.045	0.034	0.034	0.014	0.018
3	0.024	0.072	0.055	0.054	0.024	0.032
4	0.037	0.103	0.078	0.077	0.037	0.049
5	0.052	0.137	0.104	0.102	0.052	0.069
6	0.07	0.174	0.132	0.13	0.07	0.092
7	0.09	0.214	0.163	0.16	0.09	0.119
8	0.113	0.257	0.196	0.193	0.113	0.148
9	0.138	0.303	0.232	0.229	0.138	0.181

wind speed $\omega_M = v/L$. The forward stroke period T is determined from the equaltion $T = 2\beta_M/\omega_M$, where $\beta_M = \pi/18$. It follows that the time t, corresponding to the rotation for $\Delta\varphi_i$, amounts to $t = T/10$. Now, using Eqs. (6.58, 6.61), the speeds of forward movement of the corresponding actuators can be determined. For example, the speed of the actuator 6 is $v_6 = \Delta l_6/t$. Similarly, the speed of the other actuators is identified. Table 6.4 shows the results of calculations of the values of actuators axial displacement (in m.) with WB rotation angles of 0–9° and with one wind direction ($\alpha = 0$). In general, the calculations were dome for the angles of α 0° to 324° with a step of 36°.

Dependence of the displacement in the actuator at the angles of the mast rotation of 0° to 9° with the step of 1° Table 6.4.

These calculations show that with a given wind direction passing through the node C_2 as shown in Table 6.4, the largest movements are observed in the actuators 4, 5, 6. If the wind direction is changed, the greatest movements are done by another group of actuators.

Chapter 7
Parallel Manipulators' Dynamics

Abstract This chapter studies parallel manipulator dynamics using known matrix methods in the local coordinate system with the help of the recurrence formulas. The chapter discusses OPM and MPM power analysis methods based on Newton-Euler equations in the local coordinate system.

Keywords OPM and MPM power analysis · Local systems · Recurrence formulas

When studying the dynamics of the parallel manipulators, the known direct and inverse dynamics problems are solved. With respect to OPM and MPM, the range of problems to be solved is expanded, in addition, certain requirements are made to the solutions of the dynamics problems associated with the need to control the relative movement of the links. In the solution of the direct problem, reasoning from that the motion is given and corresponding to the programmed motion, efforts in the drives are determined, and efforts of the drives are required to perform the predetermined motion. The solution of this problem allows the manipulators' design to determine the optimum dimensions of the links accounting the strength and rigidity, and to choose the engines of executive drives. When calculating the actual manipulator, the direct problem also allows verifying their compliance to the set requirements for the drive engines. The inverse dynamics problem is that the known mass-inertial characteristics are determined by the movement under the influence of the predetermined loads. The resulting law of motion must conform to the required movement. However, in reality, there is not an exact match observed, since dynamic errors occur, which must be minimized by calculation and automatic motion control. The inverse dynamics problem is reduced to drawing up the most appropriate dynamic model for the actual design of a multi-degree-of-freedom mechanism. To solve the above dynamic problems, the researchers used a variety of methods. Thus, for example, the power analysis uses d'Alembert's principle [51, 52], various forms of Newton-Euler equations are applied [53–55], along with general theorems of the rigid body dynamics [56, 57]. Among the kinetostatic analysis methods, the most effective method of calculation is the recurrence formulas in the local coordinate system related to the link in question. Due to the fact that the moments of inertia and the position of the gravity center do

K. S. Sholanov, *Parallel Manipulators of Robots*, Mechanisms and Machine Science 92,
https://doi.org/10.1007/978-3-030-56073-7_7

not depend on the mechanism configuration, the said method allows implementing the algorithm of calculating the effort of the drives in real time without delay.

In addition, to output the differential equations of motion, the Lagrange equations of the second kind [58], the Euler-Lagrange equation [46], are applied. Use of the homogeneous transformation matrices in the output of the differential equations using the Euler-Lagrange equations allows applying the obtained above results of the kinematic analysis using the homogeneous transformation matrix and, thereby, achieve unity in the methods of the kinematic and dynamic analysis. For the manipulators' dynamic analysis, various authors use the principle of the least constraint by Gauss and Hamilton stationary action. When analyzing the dynamics of manipulators with a small number of degrees of freedom, constraint graphs and dynamic graphs are used. Due to the comparative analysis of the direct solution methods and the inverse problem of the parallel manipulators' research, this study prefers the method of the power analysis in the local coordinate system by the recurrence formulas.

7.1 Power Analysis of One-Loop Parallel Manipulators

7.1.1 Power Analysis of the Connecting Link and the Gripper

Let us consider the power analysis method using Newton-Euler equations. The system being researched consists of the OPM connecting link and the gripper. In its turn, the connecting link consists of two bodies capable of translational movement relative to each other. For the system consisting of the set of these bodies, the reaction forces between the bodies of the connecting link and the gripper drive force will be the internal forces. In the design model of the power analysis (Fig. 7.1), it is assumed that the connecting link at the point $\boldsymbol{A}$ is connected to the last link of the first CFG with the help of the ball joint; and at the point $\boldsymbol{B}$, it is connected to the second CFG with the help of the ball joint with the pin. The connecting link will be connected to the coordinate system $BX_cY_cZ_c$ with the center at the point $\boldsymbol{B}$, while the axis BZ_c is directed along the axis of the connecting link. In addition, the auxiliary coordinate system $CX_{cx}Y_{cx}Z_{cx}$ is chosen; this system is related to the gripper and has its center at point $\boldsymbol{C}$, and its axes are parallel to the axes having the same name and located in the coordinate system $BX_cY_cZ_c$. The choice of the additional auxiliary coordinate system is conditioned by the necessity of determining the inertial forces, and moments from the inertia forces of the gripper in the coordinate system related directly to the gripper. In Fig. 7.1, S_c denotes the mass center of the connecting link without the gripper. It is assumed that the center of mass coincides with the point of the gripper $\boldsymbol{C}$—the gripper center. The system of bodies consisting of the gripper and the connecting link is affected by the following active forces: $\bar{G}_c = m_c\ \bar{g}$, $\bar{G}_{cx} = m_{cx}\ \bar{g}$ are gravity forces of the connecting link and the gripper, respectively; $\bar{F}$ is the total force affecting the manipulator gripper from the outside. It should be noted that the nature of the forces $\bar{F}$ may differ depending on the functions performed by the

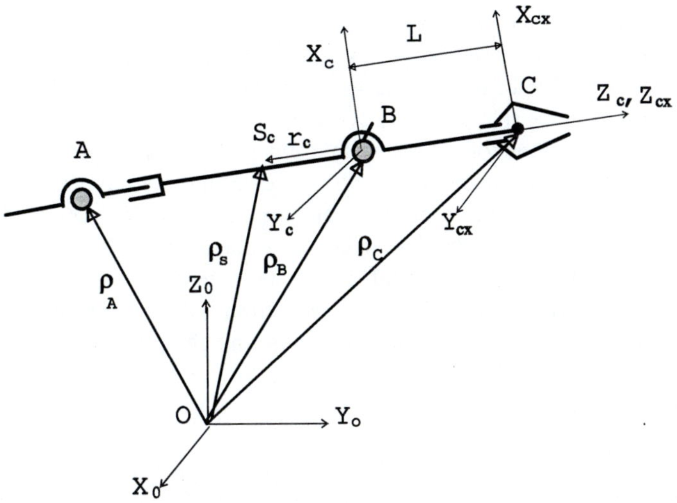

Fig. 7.1 Connecting link with the gripper

manipulator. This force and the law of its changes should be given. In our case, it is assumed that the manipulator carries a load weighing m_{Γ}, therefore, $\bar{F} = m_{\Gamma}\,\bar{g}$.

In the coordinate system $BX_cY_cZ_c$, *the* gravity forces are defined as follows:

$$G_c^c = R^{-1}\begin{pmatrix} 0 \\ 0 \\ -mg \end{pmatrix} = -m_c g \begin{pmatrix} \text{Sin}g \\ 0 \\ \text{Cos}g \end{pmatrix}. \tag{7.1}$$

The expression for the matrix R^{-1} is accepted in accordance with (5.54).

Since the axes of the coordinate systems $BX_cY_cZ_c$ and $CX_{cx}Y_{cx}Z_{cx}$ are parallel, the dependence (7.1) is obtained for the gravity forces of the gripper weighing m_{cx} and a load weighing m_{Γ}.

$$G_{\Gamma+cx} = -(m_{\Gamma} + m_{cx})g \begin{pmatrix} \text{Sin}g \\ 0 \\ \text{Cos}g \end{pmatrix}. \tag{7.2}$$

In places of the connecting link, connections to the CFG, forces, and moments act from the reaction forces. So, at the point ***B***, the reaction force $B^c = \left(X_b^c, Y_b^c, Z_b^c\right)^T$ is applied. Here and hereinafter, the upper index denotes the belonging to the coordinate system. For example, the index (c) denotes that the indicated values are determined in the system $BX_cY_cZ_c$. Since the joint ***B*** has two degrees of freedom, additionally, the reaction forces are effected by the moments of the reaction forces $M_B^c = (0, 0, M_{Bc})^T$. The presence of one component of the moment of the reaction forces is conditioned by $m_c g \cdot l_{AS} \times \text{Cos}g/EA - Z_B \cdot l_{AB}\text{Cos}g/EA$ and $+\ (m_{cx} + m_{\Gamma})g \cdot l_{AC} \cdot \text{Cos}g/EA = 0$ which are the design features of the connecting link and the special choice of the axis BZ_c direction. At the point ***A***, the reaction force $A^c = (X^c_a, Y^c_a, Z^c_a)^{\mathrm{T}}$ is applied. All the abovementioned reaction forces and moments of the reaction forces are unknown values. One of the tasks of the power analysis is to determine these unknown values. However, for the spatial system of forces in question, the number of unknown values exceeds the number of the kinetostatic equations. Consequently, the problem to be solved is statically indeterminate. It is known that the missing equation can be obtained from the condition of the strain compatibility. It is assumed that the movement of the section drawn perpendicular to the axis BZ_c, and caused by the deformation in the direction of the axis AB, equals to zero.

l_{AS}, l_{AB}, and l_{AC} are the lengths of various sections of the connecting link. If we assume that the rigidness (EA) of the connecting link along the entire length is the same, the strain compatibility condition easily provides the missing dependence for the reaction force Z_B component in the joint ***B*** in the form

$$Z_B = [m_c \cdot l_{AS} + (m_{cx} + m_\Gamma) \cdot l_{AC}] \cdot g / l_{AB}. \tag{7.3}$$

In accordance with the D'Alembert principle, the active forces, the reaction force, and inertia forces form a balanced system of forces. Therefore, to make up the balance equations, inertial forces and moments of inertia forces must be applied to the system under consideration. These values were obtained using the Newton equations for translational motion and the Euler equations for the rotational motion.

First, inertial forces and moments of inertia forces are found separately for the connecting link in the coordinate system $BX_cY_cZ_c$ and the gripper in the system $CX_{cx}Y_{cx}Z_{cz}$. The main vector of the connecting link inertia force by definition equals to

$$\bar{P}^u_c = -m_c \bar{a}_{cs}, \tag{7.4}$$

where $\bar{a}_{cs}$ is the acceleration of the connecting link mass center, which is determined by the previously obtained dependence (5.56).

The main vector of the inertial forces of the gripper carrying the load is defined similar to (7.4) from the dependence

$$\bar{P}^u_{cx+\Gamma} = -(m_{cx} + m_\Gamma)\bar{a}_{cx}. \tag{7.5}$$

Wherein the acceleration of the mass center of the gripper $\bar{a}_{cx}$ is found by formula (5.56). To obtain the kinetostatic equation, the main vectors of the forces of inertia come to the point ***B*** and are recorded in the projections on the axis $BX_cY_cZ_c$. In this case, the main vector of the forces of inertia of the connecting link in the coordinate system $BX_cY_cZ_c$ looks as follows:

$$\left(\bar{P}^u_c\right)^c = -m_{cx} \begin{pmatrix} a^c_{xc} \\ a^c_{yc} \\ a^c_{zc} \end{pmatrix}. \tag{7.6}$$

The components of matrix-vector (7.6) are proportional to the projections of acceleration of the mass center of the connecting link on the coordinate axis $BX_cY_cZ_c$, which in their turn are defined as follows:

$$\begin{pmatrix} a_{xc}^c \\ a_{yc}^c \\ a_{zc}^c \end{pmatrix} = R^{-1}\bar{a}_{cs}. \tag{7.7}$$

In the right part of the matrix Eq. (7.7), the rotation matrix R^{-1} from (5.54) and column vector of acceleration of the mass center of the connecting link should be substituted in the reference coordinate system from (5.63). Similarly, the principal vector of the forces of inertia are obtained for the gripper and the load in the system $BX_cY_cZ_c$

$$\left(\bar{P}_{cx+\Gamma}^u\right)^c = -(m_{cx} + m_\Gamma)R^{-1}\bar{a}_{cx} = -(m_{cx} + m_\Gamma)\begin{pmatrix} (a_{cx})_x^c \\ (a_{cx})_y^c \\ (a_{cx})_z^c \end{pmatrix}, \tag{7.8}$$

where the column vector elements are the projections of acceleration of the gripper center on the axis of the coordinate system $BX_cY_cZ_c$, and are defined from the equation that is similar to (7.7), as the same name axis of the coordinate systems $CX_{cx}Y_{cx}Z_{cx}$ and $BX_cY_cZ_c$ are parallel to each other. The main moment of the forces of inertia is also determined separately for the connecting link and the gripper with the load. Thus, the main moment of forces of inertia of the connecting link relative to the point $\boldsymbol{B}$ in the reference coordinate system is as follows:

$$\bar{M}_{CB}^u = -(I_c\bar{e}_c + \bar{w}_c \times I_c\bar{w}_c + m_c\bar{r}_{sc} \times \bar{a}_B), \tag{7.9}$$

where $\bar{r}_{sc}$ is the column vector composed of the projections of the mass center on the axis of the reference coordinate system; $\bar{a}_B$ is the acceleration of the point $\boldsymbol{B}$ of the last CFG link determined according to (5.63). The expression (7.9) includes I_c—inertia tensor of the connecting link at the point $\boldsymbol{B}$, while the direction of the axis is parallel to the axes of the reference coordinate system. The components of the inertia tensor in this case are variable values that depend on the position of the connecting link, which greatly complicates the problem solution. Conversely, the inertia tensor of the connecting link at the center of the coordinate system $BX_cY_cZ_c$ that is invariably related to the connecting link is a predetermined mass-inertia feature of the connecting link. This fact was the main reason that the further calculations are carried out in the local coordinate system $BX_cY_cZ_c$. Thus, the inertia tensor of the connecting link in the local coordinate system is characterized by the matrix

$$I_c = \begin{pmatrix} I_x^c & -J_{xy}^c & -J_{xz}^c \\ -J_{yx}^c & J_y^c & -J_{yz}^c \\ -J_{zx}^c & -J_{zy}^c & J_z^c \end{pmatrix}, \tag{7.10}$$

where J_x^c, J_y^c, J_z^c is the axial moments of inertia of the connecting link relative to the axes of the coordinate system $BX_cY_cZ_c$; *and* $J_{xy}^c, J_{yz}^c, J_{xz}^c$ are centrifugal moments of inertia. Axis BZ_c can be taken as the main central axis of inertia, then $J_{yz}^c = J_{xz}^c = 0$. The coordinates of the mass center in this case are $x_s^c = y_s^c = 0$. The moment of inertia of the connecting link relative to the axes of the coordinate system $BX_cY_cZ_c$ can be obtained by expressing the matrix Eq. (7.9) via the projections of the corresponding values to the axes of the coordinate system $BX_cY_cZ_c$

$$\begin{aligned}
\left(M_{c\beta}^u\right)_x^c &= -J_x^c e_{xc}^c + J_{xy}^c e_{yc}^c + J_{xz}^c e_{cz}^c - J_{xy}^c w_{zc}^c w_{xc}^c + J_{xz}^c w_{xc}^c w_{yc}^c \\
&\quad - J_{yz}^c (w_{zc}^c - w_{yc}^c) - w_{zc}^c w_{yc}^c \left(J_z^c - J_y^c\right) - m_c\left(y_s a_{\beta z} - z_s a_{\beta y}\right), \\
\left(M_{c\beta}^u\right)_y^c &= J_{xy}^c e_{xc}^c - J_y^c e_{yc}^c + J_{yz}^c e_{cz}^c - J_{yz}^c w_{xc}^c w_{yc}^c + J_{xy}^c w_{zc}^c w_{yc}^c; \\
&\quad - J_{xz}^c \left(w_{xc}^c - w_{zc}^c\right) - w_{zc}^c w_{xc}^c \left(J_x^c - J_z^c\right) - m_c\left(z_s a_{\beta x} - x_s a_{\beta z}\right)^c, \\
\left(M_{c\beta}^u\right)_z^c &= J_{xz}^c e_{xc}^c + J_{yz}^c e_{yc}^c - J_z^c e_{cz}^c - J_{xz}^c w_{zc}^c w_{yc}^c + J_{yz}^c w_{xc}^c w_{zc}^c \\
&\quad - J_{xy}^c \left(w_{yc}^c - w_{xc}^c\right) - w_{xc}^c w_{yc}^c \left(J_y^c - J_x^c\right) - m_c\left(x_s a_{\beta y} - y_s a_{\beta x}\right)^c.
\end{aligned} \tag{7.11}$$

Projections of the vector product to the axis of the coordinate system $BX_cY_cZ_c$ are determined using the rotation matrix R^{-1} obtained from the matrix T^{-1}(5.54). In this case

$$\begin{pmatrix} \left(y_s a_{\beta z} - z_s a_{\beta y}\right)^c \\ \left(z_s a_{\beta x} - x_s a_{\beta z}\right)^c \\ \left(x_s a_{\beta y} - y_s a_{\beta x}\right)^c \end{pmatrix} = R^{-1} \begin{pmatrix} y_s a_{\beta z} - z_s a_{\beta y} \\ z_s a_{\beta x} - x_s a_{\beta z} \\ x_s a_{\beta y} - y_s a_{\beta x} \end{pmatrix}. \tag{7.12}$$

In Eq. (7.12), x_s, y_s, and z_s are the coordinates of the mass center of the connecting link in the reference coordinate system; a_{Bx}, a_{By}, a_{Bz} are the projections of acceleration of the point ***B*** of the CFG to the axes of the reference coordinate system. In the expression (7.11) $w_{xc}^c, w_{yc}^c, w_{zc}^c$ denote the projections of the angular velocity of the connecting link on the axes of the coordinate system $BX_cY_cZ_c$, and $e_{xc}^c, e_{yc}^c, e_{zc}^c$ are the projections of the angular acceleration of the connecting link on the axes of the same coordinate system.

While the projections of the angular velocity and acceleration in (7.11) are found from the equations

$$\begin{pmatrix} w_{xc}^c \\ w_{yc}^c \\ w_{zc}^c \end{pmatrix} = R^{-1} \begin{pmatrix} w_{xc} \\ w_{yc} \\ w_{zc} \end{pmatrix}, \quad \begin{pmatrix} e_{xc}^c \\ e_{yc}^c \\ e_{zc}^c \end{pmatrix} = R^{-1} \begin{pmatrix} e_{xc} \\ e_{yc} \\ e_{zc} \end{pmatrix}, \tag{7.13}$$

where values w_{xc}, w_{yc}, w_{zc}, e_{xc}, e_{yc}, e_{zc} are found by formula (5.37).

The main vector of the inertial forces of the gripper M_{cx}^u relative to the point ***C***—the gripper center in the reference coordinate system is conditioned by the expression similar to (7.9), in more detail

$$\bar{M}^{u}_{cx} = -(I_{cx}e_{cx} + w_{cx} \times I_{cx}\bar{w}_{cx} + (m_{cx} + m_{\Gamma})\bar{r}_{c} \times \bar{a}_{c}), \tag{7.14}$$

where I_{cx} is the inertia tensor of the gripper at the point $\boldsymbol{C}$, components of which are calculated relative to the axes of the reference coordinate system; $\bar{r}_{-} = (x_c, y_c, z_c)^T$ is the column vector composed of the coordinates of the gripper center in the reference coordinate system. If the gripper center $\boldsymbol{C}$ is the center of the coordinate system $CX_{cx}Y_{cx}Z_{cx}$, invariably related to the gripper, the mass-inertia characteristics of the gripper with the load inertia tensor is characterized by the following inertia tensor I^{cx}_{cx} at the point $\boldsymbol{C}$:

$$I^{-x}_{-x} = \begin{pmatrix} I^{cx}_{x} & -J^{cx}_{xy} & -J^{cx}_{xz} \\ -J^{cx}_{yx} & J^{cx}_{y} & -J^{cx}_{yz} \\ -J^{cx}_{zx} & -J^{cx}_{zy} & J^{cx}_{z} \end{pmatrix}, \tag{7.15}$$

where $J^{cx}_{x}, J^{cx}_{y}, J^{cx}_{z}$—are the axial moments of inertia of the gripper with the load relative to the axes of the coordinate system $X_{cx}Y_{cx}Z_{cx}$; J^{cx}_{xy}, J^{cx}_{yz}, J^{cx}_{xz}—are centrifugal moments of inertia. If the local coordinate system coincides with the main central axes, then $J^{cx}_{xy} = J^{cx}_{yz} = J^{cx}_{xz}$. Coordinates of the mass center of the gripper with the load in the local coordinate system with the center at the mass center equal to zero is $x^{cx}_{c} = y^{cx}_{c} = z^{cx}_{c} = 0$.

The moments from the inertia forces of the gripper with the load relative to the main central axis can be determined by the following formulas.

$$\left(M^{u}_{cx}\right)^{cx}_{x} = -J^{cx}_{x}(e_{cx})^{cx}_{x} - (w_{cx})^{cx}_{z}(w_{cx})^{cx}_{y}\left(J^{cx}_{z} - J^{cx}_{y}\right),$$

$$\left(M^{u}_{cx}\right)^{cx}_{y} = -J^{cx}_{y}(e_{cx})^{cx}_{y} - (w_{cx})^{cx}_{z}(w_{cx})^{cx}_{x}\left(J^{cx}_{x} - J^{cx}_{z}\right),$$
$$\left(M^{u}_{cx}\right)^{cx}_{z} = -J^{cx}_{z}(e_{cx})^{cx}_{z}. \tag{7.16}$$

By Poinsot method, the main vector of the forces applied to the gripper is brought to the point $\boldsymbol{B}$. Then, the total moment of inertia forces and the moment resulting from bringing the forces in the coordinate system of the connecting link is equal to

$$\left(\bar{M}^{u}_{cx}\right)^{c} = B\bar{C} \times \left(\bar{P}^{u}_{cx+r}\right)^{c} + E\left(M^{u}_{cx}\right)^{cx}. \tag{7.17}$$

Here, the column vector is $\bar{B}\bar{C} = (0\ 0\ L)^{\mathrm{T}}$; L is the gripper lever; E is the unit matrix representing the transformation matrix of the coordinate system $CX_{cx}Y_{cx}Z_{cx}$ into $BX_cY_cZ_c$, same name axes of which are parallel. In the projections on the axes of the coordinate system $BX_cY_cZ_c$, the presented moments of inertia are equal to

$$\left(M^{u}_{cx}\right)^{c}_{x} = -L\left(P^{u}_{cx+r}\right)^{c}_{x} + \left(M^{u}_{cx}\right)^{cx}_{x},$$
$$\left(M^{u}_{cx}\right)^{c}_{y} = L\left(P^{u}_{cx+r}\right)^{c}_{y} + \left(M^{u}_{cx}\right)^{cx}_{y},$$
$$\left(M^{u}_{cx}\right)^{c}_{z} = \left(M^{u}_{cx}\right)^{c}_{z}. \tag{7.18}$$

According to the D'Alembert principle, for the bodies' system consisting of the gripper and the connecting link, the following kinetostatic equations can be obtained in the vector form:

$$\begin{aligned}&\bar{G}_c^c + \bar{G}_{cx+\Gamma}^c + \left(\bar{P}_c^u\right)^c + \left(\bar{P}_{cx+\Gamma}^u\right)^c + \bar{A}^c + B^c = 0\\&\bar{r} \times \bar{G}_c^c + \overline{BC} \times \bar{G}_{cx+\Gamma}^c + \left(\bar{M}_c^u\right)^c + \left(\bar{M}_{cx}^u\right)^c + \bar{M}_\beta^u + \overline{BA} \times \bar{A}^c = 0.\end{aligned} \quad (7.19)$$

In Eq. (7.19), $\bar{r}_c = (0 \quad 0 \quad - BS_c)^T$, $\overline{BC} = (0\,0 - s_r)^{\mathrm{T}}$. Component BS_c denotes the distance from the point $\boldsymbol{B}$ to the mass center of the connecting link, without the gripper; s_r is the variable that is equal to the distance between points $\boldsymbol{A}$ and $\boldsymbol{B}$ of the connection of the end links of the CFG with the connecting link. This distance is defined by formula (5.24). From Eq. (7.19), algebraic kinetostatic equations can be obtained for the connecting link with the gripper in the form

$$\begin{aligned}&-(m_c + m_{cx} + m_r)g\mathrm{Sin}g - m_c a_{xc}^c - (m_{cx} + m_r)(a_{cx})_x^c + X_A^c + X_B^c = 0,\\&-m_c a_{yc}^c - (m_{cx} + m_r)(a_{cx})_y^c + Y_A^c + Y_B^c = 0,\\&-(m_c + m_{cx} + m_r)g\mathrm{Cos}g - (m_c + m_r)(a_{cx})_z^c - m_c a_{zc}^c + Z_A^c + Z_B^c = 0,\\&\left(M_c^u\right)_x^c + \left(M_{cx}^u\right)_x^c + s_r Y_A^c = 0,\ BS_c m_c g\mathrm{Sin}g - L(m_{cx} + m_r)\mathrm{Sin}g\\&+ \left(M_c^u\right)_y^c + \left(M_{cx}^u\right)_y^c - s_r X_A^c = 0,\ \left(M_c^u\right)_Z^c + \left(M_{cx}^u\right)_Z^c + M_Z^B = 0.\end{aligned} \quad (7.20)$$

The solution of the resulting system of equations considering Eq. (7.3) allows finding the unknown reaction forces A^c and B^c in the kinematic pairs $\boldsymbol{A}$ and $\boldsymbol{B}$, and the reaction moment M_z^B that limits rotation of the connecting link around its own axis. These forces and the moment are initial ones in the forces calculation for the CFG and in defining the efforts of the drives of multi-degree-of-freedom mechanisms.

7.1.2 Power Analysis of CFG by the Recurrence Formulas

To solve the problems of the CFG power analysis, let us derive the recurrent dependences.

As shown in the design model (Fig. 7.2), the position of the points O_{i-1}, O_i, which are the centers of the coordinate systems related to, respectively, I, $i+1$ links, relative to the fixed coordinate system is defined by the radius vectors $\bar{r}_{i-1}$, $\bar{r}_i$.

Radius vector $\bar{r}_{si}$ determines the position of S_i—the mass center of the i-th link in the reference system. Vectors $\bar{r}_i$, $\overline{O_{i-1}S_i}$, $\overline{O_iS_i}$ characterize the position of the center of the i-th coordinate system relative to the i–1-th one, and the position of the mass center S_i relative to the i–1 and i-th coordinate system. At the point O_{i-1}, the reaction force $\bar{F}_i$ is applied from the action of the i–1-th link to the i-th link. $\bar{M}_i$ denotes the main vector of the moments of reaction forces of the i–1 link, which act on the i-th link. At the point **K**, the force developed by the drive P_i^{np} acts. It should be noted that in this case, in contrast to what is accepted in the studies [47], from the totality

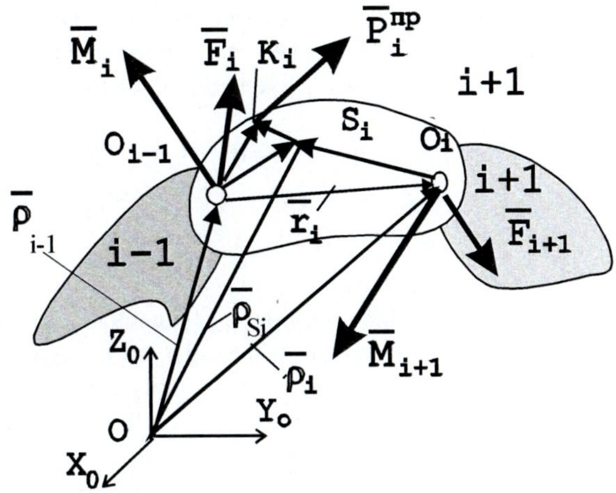

Fig. 7.2 Design model

of forces acting on the i-th link, the efforts are singled out separately, which are produced by the translational motion drives. This is related to the fact that in reality, there are manipulators, in which the drives are not integrated into the joints. In the case, when the movement in the joint is performed by the rotational motion drives, the moment transmitted by the drive is regarded as a component of the moment of the reaction forces $\bar{M}_i$ in the articulation joint. Since the i-th link is considered as a system of bodies, the total gravity force $\bar{G}_i$ (including the drive weight) must be applied in the mass center of the system S_i. Let the total force of $\bar{F}_{i+1}$ and the force moment of $\bar{M}_{i+1}$ be acting on the i + 1-th link from the side of the i-th link. Then, according to the third Newton law, the i-th link from the side of the i + 1 side is effected by the force equal to $(-\bar{F}_{i+1})$ and the moment of $(-\bar{M}_{i+1})$. In accordance with the mass center theorem for the i-th link, it follows that

$$m_i \bar{a}_{si} = \bar{F}_i - \bar{F}_{i+1} + \bar{P}_i^{\bullet p} + \bar{G}_i, \tag{7.21}$$

where m_i and a_{si} are the weight and acceleration of the mass center of the i-th link.

According to the kinematic moment change theorem, relative to the fixed point O recorded for the i-th link, the following vector equation is performed:

$$\begin{aligned} I_i \bar{e}_i + \bar{w}_i \times (I_i \bar{w}_i) &= \bar{M}_i - \bar{M}_{i+1} + (\bar{r}_{i-1} - \bar{r}_{si}) \times \bar{F}_i - (\bar{r}_i - \bar{r}_{si}) \\ &\times \bar{F}_{i+1} + \overline{S_i K_i} \times \bar{P}_i^{np} = \bar{M}_i - \bar{M}_{i+1} + (\bar{r}_{i-1} - \bar{r}_{si}) \\ &\times \left(m_i \bar{a}_{si} - \bar{G}_i\right) + \overline{O_{i-1} K_i} \times \bar{P}_i^{np} - \bar{r}_i \times \bar{F}_{i+1}. \end{aligned} \tag{7.22}$$

In the expression (7.21), I_i is the inertia tensor of the i-th body at the point O. The position of the point of the force application from the drive $P_i^{\bullet p}$ relative to the mass center $\overline{O_{i-1} K_i}$ is determined from the vector equation

$$\overline{S_i K_i} = \overline{S_i O_{i-1}} + \overline{O_{i-1} K_i} = \left(\bar{r}_{i-1} - \bar{r}_{si} + \overline{O_{i-1} K_i}\right). \tag{7.23}$$

It should be noted that in the case, when the drive effort is applied in the kinematic pair ($\overline{S_i K_i} = 0$), $P_i^{\bullet p}$ equals to the reaction forces component along the axis $O_{i-1}Z_{i-1}$. Expression (7.21) can be represented in the recurrent form

$$\begin{aligned}\bar{F}_i &= m_i \bar{a}_{si} + \bar{F}_{i+1} - \bar{P}_i^{\bullet p} - \bar{G}_i, \quad \bar{M}_i = \bar{M}_{i+1} + \left(\bar{r}_i + \overline{O_i S_i}\right) \\ &\times \left(m_i \bar{a}_{si} - \bar{G}_i\right) - \overline{O_{i-1}K_i} \times \bar{P}_i^{np} + \bar{r}_i \times \bar{F}_{i+1} + \\ &+ I_i \bar{e}_i + \bar{w}_i \times (I_i w_i).\end{aligned} \tag{7.24}$$

Equality (7.23) allows computing sequentially, starting with the last link, $\bar{F}_i$, $\bar{M}_i$ with $\bar{F}_{i+1}$, $\bar{M}_{i+1}$, derived from the calculations given in the previous step. Power calculation starts from the last link, in this case, assuming the $\bar{F}_{i+1}$, $\bar{M}_{i+1}$ equals to the loads acting on the gripping tool or on the part of the connecting link.

However, if you follow the above method of the kinetostatic calculation, then certain difficulties may occur in respect to primarily the fact that the equality (7.23) requires cumbersome calculations to be done. Indeed, existing in (7.23) values of the elements of the inertia tensor I_i vary with the change of the position of the i-th link. This requires, while changing the link position, to perform each time additional calculations of the elements of the inertia tensor I_i. For this reason, further values included in Eq. (7.23) are expressed in terms of their values in the local coordinate system related to the i-th link. Just as kinematic parameters of the CFG were obtained in the system of coordinates of the link (paragraph 6.2), the recurrent dependences (7.23) can also be represented in the local coordinate system of the link. So, using the submatrix R_i^{i-1}, R_{i-1}^i(5.4), it is possible based on (7.23) to obtain the recurrent dependences for the power analysis, and record them in the local coordinate system related to the i-th link, as follows:

$$\begin{aligned}R_0^i \bar{F}_i &= R_{i+1}^i\left(R_0^{i+1}\bar{F}_{i+1}\right) + m_i R_0^i \bar{a}_i - R_0^i\left(\bar{G}_i + \bar{P}_i^{\bullet p}\right) \\ R_0^i \bar{M}_i &= R_{i+1}^i\left[R_0^{i+1}\bar{M}_{i+1} + \left(R_0^{i+1}\bar{r}_i\right) \times \left(R_0^{i+1}\bar{F}_{i+1}\right)\right] \\ &+ \left(R_0^i \bar{r}_i + R_0^i \overline{O_i S_i}\right) \times \left(R_0^i m_i a_i - R_0^i \bar{G}_i\right) + R_0^i P_i^{\bullet p} \times R_0^i \overline{O_{i-1}K_i} \\ &+ \left(R_0^i I_i R_i^0\right)\left(R_0^i \bar{e}_i + \left(R_0^i \bar{w}_i\right) \times \left[\left(R_0^i I_i R_i^0\right)\left(R_0^i \bar{w}_i\right)\right]\right.\end{aligned} \tag{7.25}$$

where ($R_0^i \overline{O_i S_i}$) is the column matrix consisting of the coordinates of the mass centers of the i-th link in the i-th coordinate system. The components of this matrix do not depend on the current mechanism configuration. Column vector composed of the coordinates of the center of the i–1 system, relative to the i-th coordinate system is defined as follows:

$$R_0^i \bar{r}_i = \begin{pmatrix} a_i \\ s_i \mathrm{Sin} a_i \\ s_i \mathrm{Cos} a_i \end{pmatrix} \tag{7.26}$$

The elements of this matrix are constant, and are input at the stage of choosing the coordinate systems. The calculations are considerably simplified by that in the matrix Eq. (7.25), the expression $(R_o^i I_i R_i^0)$ is the inertia tensor of the i-th body in the mass center, i.e.,

$$\left(R_o^i I_i R_i^0\right) = \begin{pmatrix} I_x & -J_{xy} & -J_{xz} \\ -J_{yx} & J_y & -J_{yz} \\ -J_{zx} & -J_{zy} & J_z \end{pmatrix}, \tag{7.27}$$

where the components of the inertia tensor are constant, axial and centrifugal moment of inertia relative to the axes passing through the mass center of the i-th link are parallel to the same name axis of the i-th coordinate system.

One of the tasks of the power calculation is to determine the force required for the motion of the i-th link with respect to the i–1-th link provided by drives. In this case, if the i-th kinematic pair is rotational, the moment produced by the drive equals to

$$M_i^{np} = \left(R_0^i \bar{M}_i\right)\left(R_{i-1}^i \bar{z}_0\right) + b_i \dot{q}_i. \tag{7.28}$$

In the case, when the i-th kinematic pair is translational, and $\overline{S_i K_i} = 0$, the drive effort is equal to

$$P_i^{np} = \left(R_0^i \bar{F}_i\right)\left(R_{i-1}^i \bar{z}_o\right) + b_i \dot{q}_i. \tag{7.29}$$

In these formulas, $\bar{z}_0 = (0 \;\; 0 \;\; 1)^{\mathrm{T}}$ is axis cross heading; b_i is the coefficient of viscous friction in the kinematic connection. When using the description of the interconnection of two links by the transformation matrix of eight parameters (paragraph 4.1), at the step of the design calculations, it becomes possible to select the center of the coordinate system at the mass center of the body. In this case, the recurrent formulas are considerably simplified. Indeed, if the point S_i coincides with O_i (Fig. 7.2), then there is the expression $R_o^i \overline{O_i S} = 0$ in Eq. (7.24). All off-diagonal components of the inertia tensor (7.25) will also equal to zero. In addition, the expressions for the kinematic parameters will also be simplified. The rotation matrices used in this case will be submatrices (paragraph 4.1), i.e., when describing the interconnection by the method proposed in this paper, calculations according to the recurrent formulas are simplified. Thus, the algorithm is obtained for the power analysis of the CFG multi-degree-of-freedom mechanisms according to the recurrent formulas in the local coordinate system related to the link. Recurrent dependences can be used in different methods of the mathematical description of the two bodies interconnection. According to the calculation results, effort provided by the drives is calculated, as well as the forces and the moments of the reaction forces in the kinematic pairs. The effectiveness of a particular algorithm largely depends on the speed of computation. In this connection, it should be noted [87] that the time spent for calculations using the Newton-Euler equations is twice shorter than the time spent for the calculations using Lagrange equations of the second kind.

7.2 Power Analysis of the Multiloop Manipulator

As shown above, the manipulator dynamic model can be built based on the Newton-Euler equations. Constructed dynamic model is usually used when forming the dynamics of the control systems of the executive drives of robots. For multiloop manipulator, construction of the dynamic model is a difficult task. This is seen at least from the fact that the solution of the links position problem for the multi-loop manipulator, described in paragraph 6.1, is time consuming and depends on the configuration variants. In practice, usually during the front-end engineering, it is obligatory to calculate and select the drive. Only then the dynamic model is formed for the synthesis of this drive control system. Below, there is the power analysis by the kinetostatic method, which is required for calculating and selecting executive drives of robots. The problem is solved with the following wording. Known values: mass characteristics of the moving system that is rigidly connected to the upper platform of the manipulator, the law of motion of the moving system, and the forces acting on this system. As a result of solving the problem, the forces occurring between the drives and the moving platform are determined. These efforts are subsequently used to select the drives. The initial data used are the main vector of the external forces and inertia forces $\vec{\mathbf{R}}$, which is brought to the point O_1 of the moving platform (Fig. 7.3) and the main vector of the moment of the external forces and the forces of inertia $\vec{\mathbf{M}}$. Besides the specified forces, the platform 2 is affected by the unknown forces N_3, N_4, N_5, N_6, N_7, and N_8, which are directed along the axes of the drives. Directions of

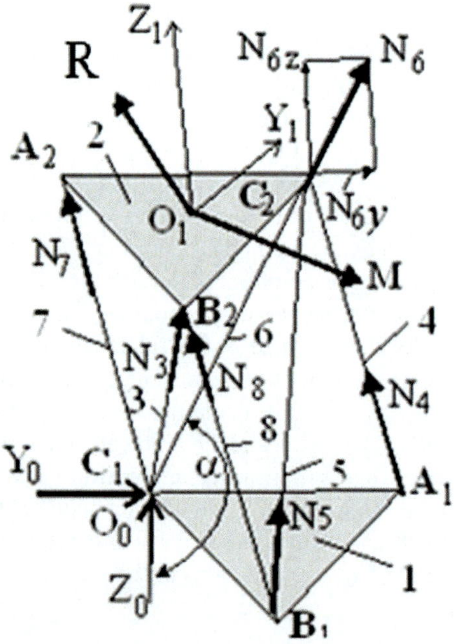

Fig. 7.3 Design model for the MPM power analysis

the lines of action of these forces for each position of the MPM are found (p. 6.3) by solving the inverse position problem. Also with the known position, the coordinates of the nodes are known, and hence the coordinates of the points of forces application are known too. Let us choose the fixed coordinate system $O_0\, X_0Y_0Z_0$ related to the bottom platform of the MPM with the center at the node C_1. Since $\bar{R}$ and $\vec{M}$ include forces of inertia, then the kinetostatic equation for the platform 2 has the following form:

$$\sum_{n=3}^{n=8} N_{nx} + R_x = 0;\ \sum_{n=3}^{n=8} N_{ny} + R_y = 0;\ \sum_{n=3}^{n=8} N_{nz} + R_z = 0;$$
$$\sum_{n=3}^{n=8} M_x(N_n) + M_{Rx} = 0;\ \sum_{n=3}^{n=8} M_y(N_n)$$
$$+ M_{Ry} = 0;\ \sum_{n=3}^{n=8} M_z(N_n) = 0. \tag{7.30}$$

Here, there are the projections of forces on the axes of the fixed coordinate system and of moments of forces relative to the axes of the fixed coordinate systems O_0 $X_0Y_0Z_0$. Figure 7.3 shows the efforts $\boldsymbol{N_i}$ acting along the axis of the drive, and the angle α between the vector of the force $\boldsymbol{N_6}$ and axis $O_0\, Z_0$ is shown as an example. Thus, in the six kinostatic equations, the unknown variables are the values of the six reaction forces—$\boldsymbol{N_i}$ ($i = 3,\ldots,8$).

Thus, the system of Eqs. (7.29) can be written as follows:

$$\begin{aligned}
&N_3\mathrm{Cos}\alpha_3 + N_4\mathrm{Cos}\alpha_4 + N_5\mathrm{Cos}\alpha_5 + N_6\mathrm{Cos}\alpha_6 + N_7\mathrm{Cos}\alpha_7\\
&\quad + N_8\mathrm{Cos}\alpha_8 + R_X = 0,\ N_3\mathrm{Cos}\beta_3 + N_4\mathrm{Cos}\beta_4 + N_5\mathrm{Cos}\beta_5\\
&\quad + N_6\mathrm{Cos}\beta_6 + N_7\mathrm{Cos}\beta_7 + N_8\mathrm{Cos}\beta_8 + R_Y = 0,\ N_3\mathrm{Cos}\gamma_3\\
&\quad + N_4\mathrm{Cos}\gamma_4 + N_5\mathrm{Cos}\gamma_5 + N_6\mathrm{Cos}\gamma_6 + N_7\mathrm{Cos}\gamma_7 + N_8\mathrm{Cos}\gamma_8\\
&\quad + R_Z = 0,\ y^0_{A2}\cdot N_7\mathrm{Cos}\gamma_7 - z^0_{A2}\cdot N_7\mathrm{Cos}\beta_7 + y^0_{B2}(N_3\mathrm{Cos}\gamma_3 + N_8\mathrm{Cos}\gamma_3)\\
&\quad - z^0_{B2}(N_3\mathrm{Cos}\beta_3 + N_8\mathrm{Cos}\beta_8) + y^0_{C2}(N_4\mathrm{Cos}\gamma_4 + N_5\mathrm{Cos}\gamma_5 + N_6\mathrm{Cos}\gamma_6)\\
&\quad - z^0_{C2}(N_4\mathrm{Cos}\beta_4 + N_5\mathrm{Cos}\beta_5 + N_6\mathrm{Cos}\beta_6) + y^0_{O1}\cdot R_Z - z^0_{O1}\cdot R_Y + M_{RX} = 0,\\
&z^0_{A2}\cdot N_7\mathrm{Cos}\alpha_7 - x^0_{A2}\cdot N_7\mathrm{Cos}\gamma_7 + z^0_{B2}(N_3\mathrm{Cos}\alpha_3 + N_8\mathrm{Cos}\alpha_3)\\
&\quad - x^0_{B2}(N_3\mathrm{Cos}\gamma_3 + N_8\mathrm{Cos}\gamma_8) + z^0_{C2}(N_4\mathrm{Cos}\alpha_4 + N_5\mathrm{Cos}\alpha_5 + N_6\mathrm{Cos}\alpha_6)\\
&\quad - x^0_{C2}(N_4\mathrm{Cos}\gamma_4 + N_5\mathrm{Cos}\gamma_5 + N_6\mathrm{Cos}\gamma_6) + x^0_{O1}\cdot R_Y\\
&\quad - z^0_{O1}\cdot R_X + M_{RY} = 0,\ x^0_{A2}\cdot N_7\mathrm{Cos}\beta_7 - y^0_{A2}\cdot N_7\mathrm{Cos}\alpha_7\\
&\quad + x^0_{B2}(N_3\mathrm{Cos}\beta_3 + N_8\mathrm{Cos}\beta_8) - y^0_{B2}(N_3\mathrm{Cos}\alpha_3\\
&\quad + N_8\mathrm{Cos}\alpha_8) + x^0_{C2}(N_4\mathrm{Cos}\beta_4 + N_5\mathrm{Cos}\beta_5 + N_6\mathrm{Cos}\beta_6)\\
&\quad - y^0_{C2}(N_4\mathrm{Cos}\alpha_4 + N_5\mathrm{Cos}\alpha_5 + N_6\mathrm{Cos}\alpha_6) = 0.
\end{aligned}$$

Table 7.1 Results of calculating the axial forces acting on the actuators

	Actuators, N					
φ (°)	No.					
	3	4	5	6	7	8
0	$-1.73*10^{-12}$	$6.113*10^3$	$2.406*10^{-12}$	946.416	$1.508*10^4$	$-0.873*10^4$
1	−0.648	$6.148*10^3$	−0.65	951.625	$1.52*10^4$	$-1.88*10^4$
2	−2.597	$6.182*10^3$	−2.617	957.264	$1.532*10^4$	$-1.887*10^4$
3	−5.858	$6.215*10^3$	−5.926	963.301	$1.544*10^4$	$-1.893*10^4$
4	−10.439	$6.247*10^3$	−10.6	969.703	$1.556*10^4$	$-1.899*10^4$
5	−16.349	$6.277*10^3$	−16.663	976.435	$1.568*10^4$	$-1.905*10^4$
6	−23.595	$6.305*10^3$	−24.137	983.46	$1.58*10^4$	$-1.91*10^4$
7	−32.184	$6.332*10^3$	−33.044	990.739	$1.591*10^4$	$-1.915*10^4$
8	−42.122	$6.357*10^3$	−43.405	998.23	$1.603*10^4$	$-1.919*10^4$
9	−53.412	$6.38*10^3$	−55.238	$1.006*10^3$	$1.614*10^4$	$-1.925*10^4$

As a result of solving system of these six equations, the sought forces N_3–N_8 affecting each of the six actuators are determined. Table 7.1 shows the data obtained as a result of calculating the axial forces acting on the actuators from the side of the manipulator's top platform when mobile platform rotation angles are 0–9° and $\alpha = 0$. Efforts of the actuators 3–8 in N depend on the angle of the mast rotation equal to 0°–9° with a step of 1°.

Analysis of the data shows that the greatest forces act on the edges of the actuators 4, 7, and 8. Therefore, in these actuators, motion with the highest power is observed.

Part IV
Use of One-Loop and Multiloop Robot Manipulators

Chapter 8
Examples of One-Loop Manipulator Applications

Abstract This chapter consists of materials related ti the use of one-loop manipulators of robots. This chapter gives some examples of the OPM application based on the author's experience. One of the OPM types in the form of six-axis table is produced as a demonstration model (KazNTU robot). This model demonstrates that it is possible to create six-axis metal-working machinery for producing details with complex configuration. With just slight modifications, KazNTU robot can be used as a mixer for multicomponent mixtures.

Keywords Demonstration sample · KazNTU robot · Six-axis table · Mixer robot

8.1 Description of Robots with One-Loop Manipulators

Above, paragraph 1.2. describes the method of obtaining one-loop manipulators and provides a series of structural diagrams made of two complete or incomplete CFG. These manipulators do not have many disadvantages that are characteristic for manipulators with the open kinematic chain, and allow the working body of the robot to perform controlled motions with six degrees of freedom. The main advantages of the OPM consist in that they have a large ratio of the load weight to the robot weight, have high rigidity, they are less powered, and provide an easy way to carry out controlled spatial motion, since in these mechanisms the direct and inverse position problems are solved unambiguously. However, there are peculiarities in the OPM application; they are the limited maintenance zone and which presence of complex connections.

To justify the functional opportunities based on the structural diagram of the manipulator of the six-axis table (Fig. 1.6), the demonstration model of the manipulator is produced and illustrated in Fig. 8.1. The OPM design uses the screw-nut mechanism for transforming the rotational motion of electric engines into the translational motion of the nut. On the fixed base 1, there are two drives with parallel axes of screws. Rotation of the screws is provided by the electric engines 2 and 3. Drives of the vertical motion also include screws with parallel axes. The engines 4.5 move the connecting link 8 in the vertical plane. The connecting link has the

K. S. Sholanov, *Parallel Manipulators of Robots*, Mechanisms and Machine Science 92,
https://doi.org/10.1007/978-3-030-56073-7_8

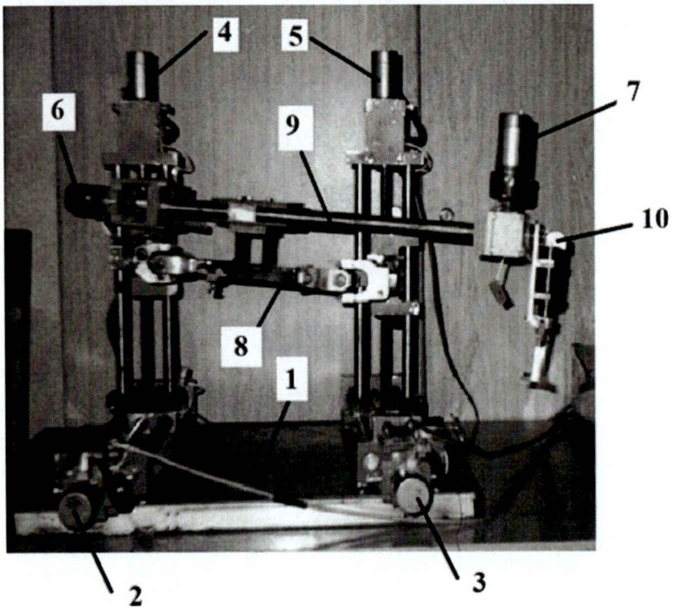

Fig. 8.1 OPM demonstration model

design represented in Fig. 1.1. Extension of the working body 10 toward the axis of the perpendicular plane, in which the axes of the vertical and horizontal screws are located, is performed by the engine 6 by moving the link 9 along the guides. This ensures the spatial orientation of the connecting link. If we take into account that the working body 10 rotates around the connecting link with the help of the engine 7, then the working body has six degrees of freedom.

Based on the demonstration model of the six-degree-of-freedom OPM, KazNTU robot is created (Fig. 8.2), in which the electric engines are equipped with tachometers providing feedback for the engine control system. Engine program control is performed by a microcontroller.

Today, six-DOF OPM are not widely used for solving technical and technological problems. At the same time, it follows that the architecture of, for example, KazNTU robot allows using it with large loads on the output link. There is no doubt that it is possible to create six-axis metal-working machinery for processing details with complex configuration. With just slight modifications, KazNTU robot can be used as a mixer for multicomponent mixtures (Fig. 8.3). It is sufficient to exclude one translational motion along the axis of the connecting link and provide the rotation of the container with the mixture relative to the axis of the connecting link. Such a mixer will perform five motion and ensure improved performance and improved quality of the mixture. OPM with the architecture of the KazNTU robot, just like other more than 20 types of the OPM, has the following advantages:

Fig. 8.2 Robot KazNTU

Fig. 8.3 Mixer for multicomponent mixtures

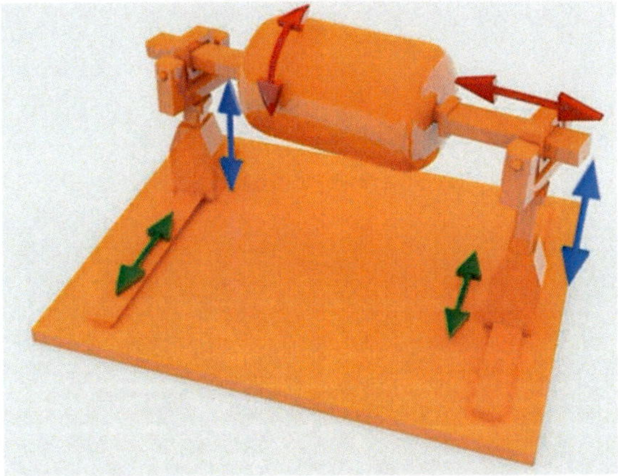

- Rigidity of its design and, as a result, accuracy in positioning and high load-carrying capacity;
- Due to the possibility of the analytic solution of the direct and inverse kinematic problem, it is easier to describe the dynamics of the system and it is possible to simulate the dynamics of the control system.

Chapter 9
Applications for Multiloop Manipulators

Abstract The chapter gives examples of the possible application of the SHOLKOR MPM. At the same time, each example of use is justified by the operative demonstration model. Examples are provided for the SHOLKOR MPM application in medicine as human cervical spine prosthesis; in seismic protection as an active support; in renewable energy sphere as a manipulator transformer for wind and wave power stations.

Keywords SHOLKOR MPM · Cervical spine prosthesis · Active support · Manipulator transformer · Waver power station · Wind power station

9.1 Human Cervical Spine Prosthesis

According to the estimates of the World Health Organization (WHO), about 13–15% of the global population have disabilities. According to the survey, which was conducted in the European Union, it was found that there are about 15% of these people in the member countries of the EU. Treatment of the spasmodic torticollis (ST) remains topical until present time, known conservative methods and surgical treatment of the ST are ineffective or accompanied by a number of complications [60]. In the population, the torticollis occurs in 1 out of 10,000 people, and about 1.5 times more often in women than in men. By its frequency, the torticollis ranks third among the congenital diseases of the locomotor apparatus (spasmodic torticollis) followed by the congenital clubfoot and congenital dislocation of the hip. In this regard, the design has been developed and an working model of the cervical spine prosthesis was produced. Moreover, a special program is developed for each patient aimed at the correction and restoration of the distorted dynamic stereotype that occurred as a result of the disease (pathological position of the head, by the forced movements of the neck muscles). There are no analogue neck prosthesis for the treatment and rehabilitation of patients with spasmodic torticollis in the world. Review of the previous researches conducted in the world regarding the research subject in question shows that to date prosthesis performing the functions of the neck muscles have not been used [61]. This is a new trend in neuro-orthopedics. The neck prosthesis allows performing the

K. S. Sholanov, *Parallel Manipulators of Robots*, Mechanisms and Machine Science 92,
https://doi.org/10.1007/978-3-030-56073-7_9

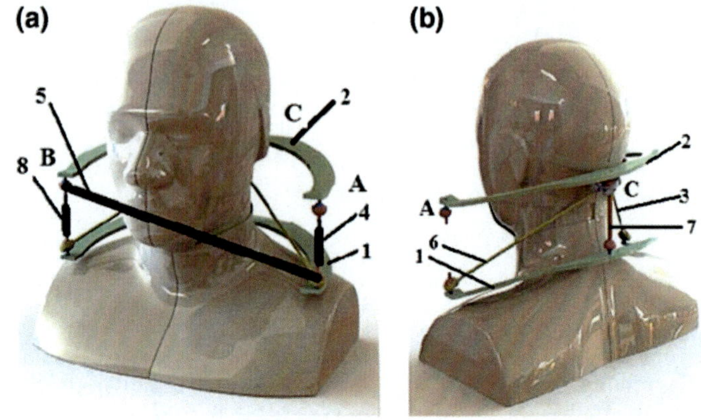

Fig. 9.1 The neck prosthesis on a Man-Simulator

angular displacement for up to 9° in the frontal and sagittal planes, thereby it partially copies the physiological movements. The proposed prosthesis provides the ability to perform all the physiological head movements, while freeing the neck from the physical exercise. The advantage of the SHOLKOR type prosthesis consists in its non-surgical fastening to the body and head. The neck prosthesis (Fig. 9.1) has a half-ring 1 connected to the frame and a half-ring 2 connected to the head. Designations of the connecting links are the same as in the SHOLKOR MPM. The difference consists in that the prosthesis has three controllable drives 4, 5, 8. The principle of the prosthesis operation is based on that the point *C* is aligned with the point in the occipital region. The position of the point *C* during installation of the prosthesis is adjusted manually by changing the lengths of the connecting links 3, 6,7. Later this point remains fixed during the head movement performed using the prosthesis. Two controlled head movements: turn and nodding are performed by changing the lengths of the connecting links 5.8, i.e., by moving the point B with the help of controlled drives.

The angle of inclination of the head to the shoulders is changed by moving the point A of the prosthesis. The upper sectional ring 2 of the prosthesis (Fig. 9.2) with three vertical plates 11 with adjustable length is rigidly attached to the plaster or plastic helmet 12 firmly fitted on the head. The lower sectional ring 1 with the shoulder piece 13, slings 14 and a belt 15 is fixed to the body of a man. The work of the controlled prosthesis is based on the functionality of the SHOLKOR MPM, as described above. Physiological movement of the head relative to the point *C* (Fig. 9.1) are approximately spherical movement and can be implemented as follows: extension and flexion (head nodding)—drive 8; lateral tilting of the head—drive 4; head turn—drive 5.

In order to investigate how the prosthesis can provide physiologically required amount of head movements, the prosthesis computer simulations was performed using MatLab software according to the algorithm provided in paragraph 6.2. The program is compiled based on the algorithm obtained by transforming the coordinate

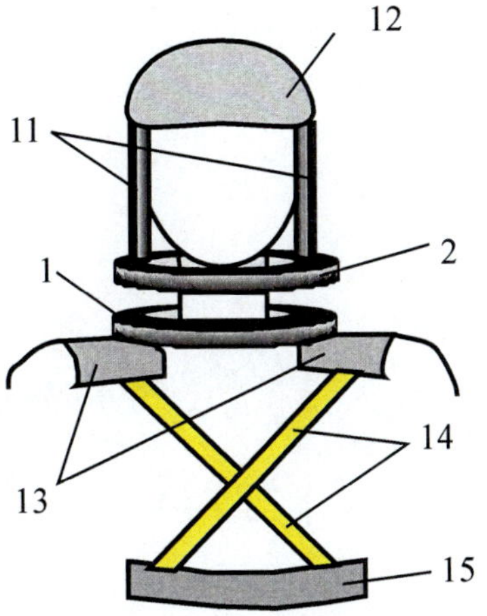

Fig. 9.2 Securing of the prosthesis

systems related to the moving links in the fixed coordinate system, and description of these transformations using the homogeneous transformation matrices. The fixed coordinate system is related to the bottom platform and directed along the side C_1A_1, and the axis C_1Z is directed perpendicular to the bottom platform plane. It is assumed that the coordinates of the node C_2 of the upper ring are known, as these coordinates are input annually depending on the anatomical features of the person.

The results of calculations in Matlab visually demonstrate the spatial position of the upper ring of the prosthesis ABC (that is rigidly connected to the head) during movements performed by each of the three drives individually (Fig. 9.3).

The analysis of virtual displacements shows that the prosthesis allows the head to perform the required volume of physiological movements. The total amount of movement in the cervical region is judged by the maximum angle of flexion of the head, its extension, lateral inclination and turn. The total amount of movements in the cervical region, which are typical for persons over 65 years, is following: angle of extension—flexion—35°, tilt—20°, turn—45°.

Based on the scheme of the prosthesis, using the Solid Work software, the prosthesis deign is developed (Fig. 9.4). The developed design has increased heelplate-shaped half-rings above and below, the curved profiles for the side connecting links formed so that they do not touch the neck.

As a result, a prototype of the prosthesis (Fig. 9.5) has been produced, which demonstrates the fact that in future it will be possible to make special medical equipment to replace the neck function using the design created on the basis of the SHOLKOR manipulator.

Three engines of the prosthesis are controlled by a manual control panel. Its power source is a battery of 12 V.

The electrical circuit shown in Fig. 9.6. allows controlling the engine speed, changing the rotation direction. The end probes stop the engine, when the displacement exceeds the limit.

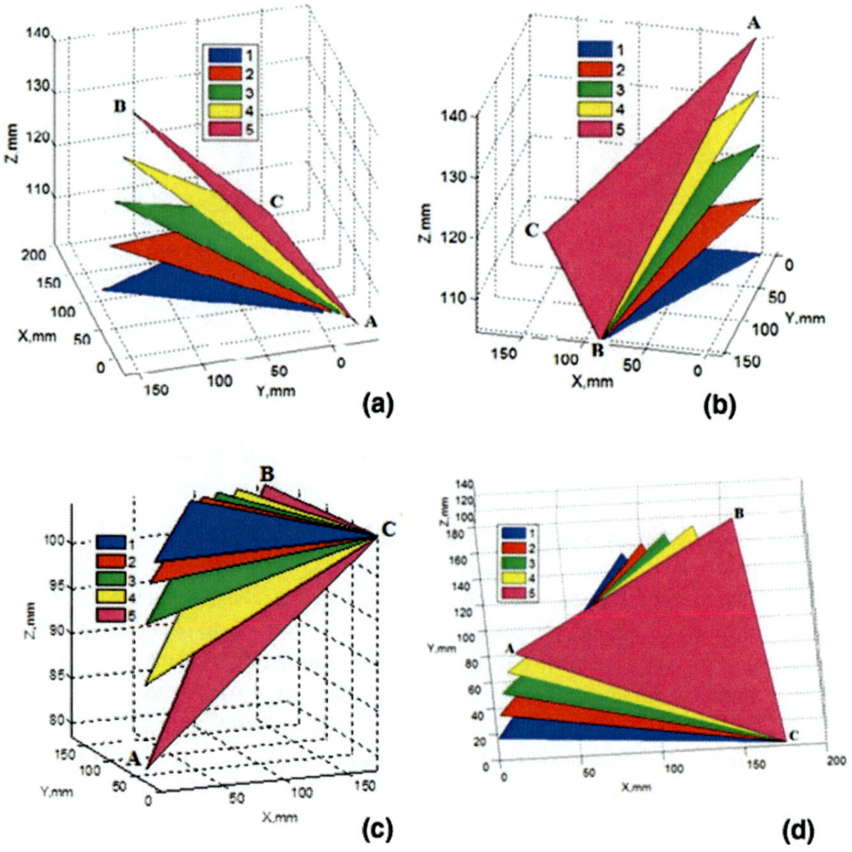

Fig. 9.3 Diagrams of visualizing the motions of the platform (Upper Ring)

9.2 Active Support Based on the SHOLKOR MPM

Currently, the various fields of activity use a lot of platform-type robots, with the design topology similar to the Gough-Stewart platforms [62–64]. The advantages of these robots are well known. However, in most platform type robots with more than 3 DOF, the reconfigurable coordinates of the moving platform are connected to each other, so that a motion of any drive requires simultaneous, coherent motion of the other drives connecting the platforms. For the manipulators of such platform type robots, there is no explicit solution of the direct kinematics problem [1–3]. In this regard, problems occur when creating the control systems, as well as during their operation, since a failure of one drive leads to the malfunction of the entire robot.

In order to improve the functionality of the platform type parallel manipulators, a new manipulator design topology is proposed; it provides a number of advantages over existing Gough-Stewart designs; and the SHOLKOR MPM is produced on this base.

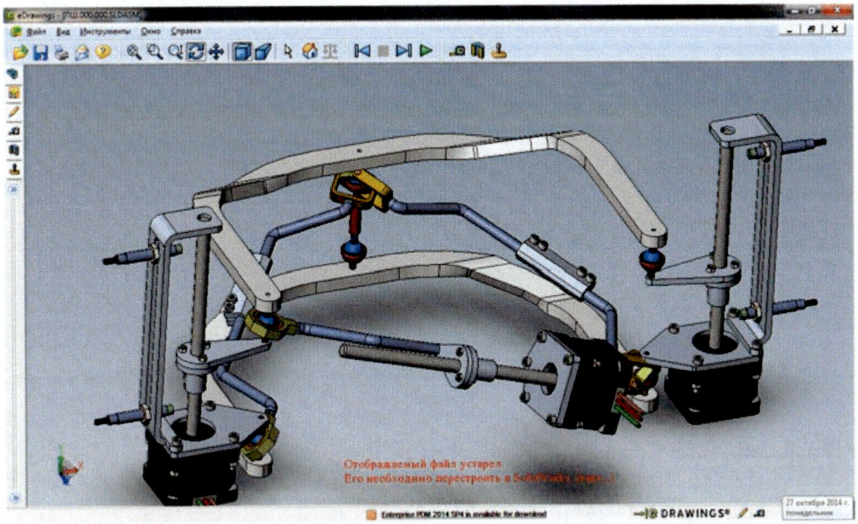

Fig. 9.4 3D model of the neck prosthesis

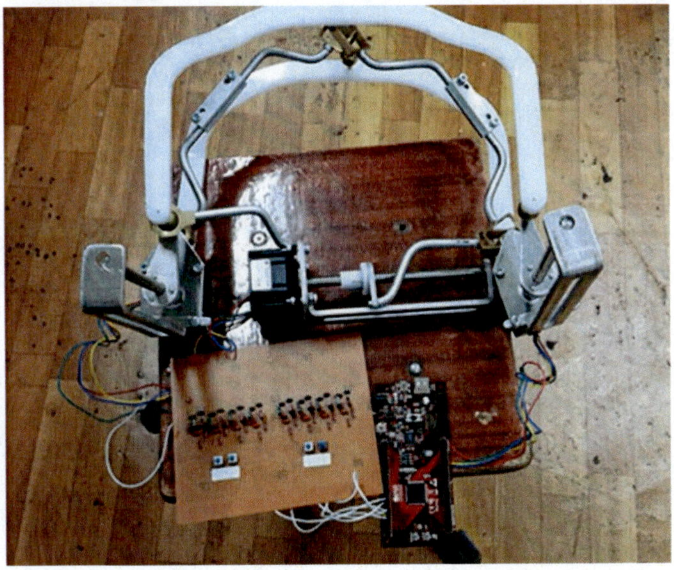

Fig. 9.5 Controlled prosthesis model

Based on the structural diagram of the manipulator (Fig. 2.8), the working model of the platform type manipulator (Fig. 9.7) was created. At the same time the connecting links are represented by unilateral controlled hydraulic drives 3; in nodes C_1 and C_2 spherical joints of 4 links are installed, and in the nodes B_1, B_2—spherical joints of 3

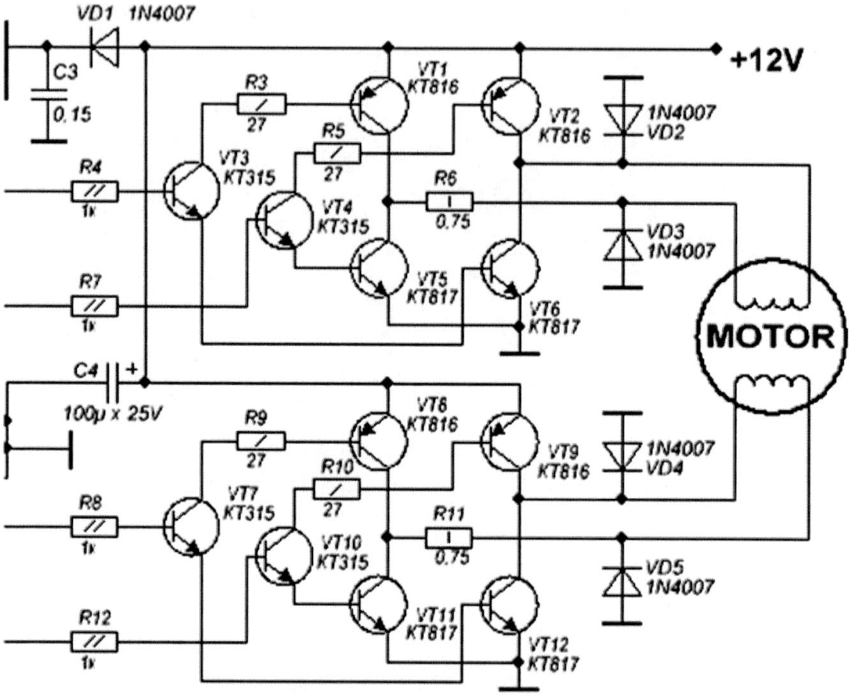

Fig. 9.6 The diagram of connecting the prosthesis drive engine

Fig. 9.7 MPM model with manual control

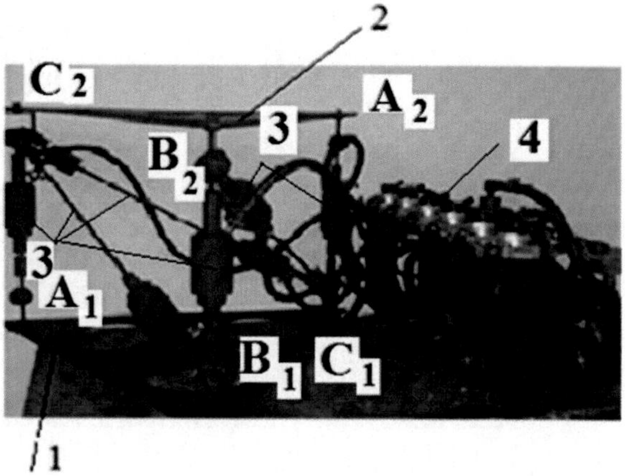

links. The functioning model of the platform type manipulator with the platforms 1 and 2 controlled by the hydraulic distributor 4, has demonstrated unique properties and possibilities conditioned by the characteristics of its design topology (paragraph 2.3.10). The properties listed above are not peculiar for any of the known platform robots with 6 degrees of freedom. Due to these properties and capabilities, the platform robot can be used as an active controlled support installed between the protected object and the source of emergency mechanical impacts. The non-standard mechanical impact is understood as the dynamic or kinematic effects, which lead to undesirable consequences. The emergency mechanical impacts can include earth crust vibrations in different planes with large amplitude during earthquakes, fluctuation of the water surface acting on the floating crafts, shocks during the landing of aircrafts and etc. The need for protection from the emergency mechanical impacts, including vibration, is especially important for vehicle drivers and passengers. Factors that negatively affect human health in vehicles are vibrations generated by the engine, transmission, suspension system and the dynamic loads associated with the crossing places with the specific type of movement, strikes, etc. The influence of these factors on human at a traffic accident [65], on tractors [66, 67], helicopters [68] cause mainly injury of the vertebral region. To develop and improve protection means, researches are carried out on human models. Of these, the most interesting are five-DOF models with 23 elements [69] and six-DOF models [70]. First studies on the vibration protection systems management include the paper, hereinafter, the active vibration isolation problems are considered in conjunction with the positioning, and separately cabin suspension parameters are chosen with pulse action due to the roughness of the road surface. There are popular studies of scientists aimed at the research of the active vibration isolation of the two-mass model and seismic protection of the object [71]. Semi-active suspension of a vehicle with magnetorheological dampers [72] and a system of 4 hydropneumatic shock absorber [73], pneumatic shock absorbers [74] using negative stiffnesses show that to improve the suspension systems of vehicles, controlled shock absorbers are used, wherein the intelligent control is used in some cases. Thus, the analysis of the studies shows that the problem of the dissipation of external force impacts is topical. There are studies aimed at the research of vibration protection and positioning of one-two–three-four-mass systems and multimass systems [75], modeling of the vehicle semi-active suspension with intelligent control, five, six, seven-DOF models [76] of a human as an object are developed, there is a variety of controlled dissipation devices. This analysis of researches published earlier, confirms that the idea of using the MPM with six degrees as active supports is significantly original. For the scientific substantiation of using the SHOLKOR MPM as a six-axis active supports, its functionality is studied. For the use as an active support and to provide the great weight carrying capacity, the platform robot uses automatically controlled hydraulic drives (controlled shock absorbers).

Mechanism of quenching the emergency mechanical impacts is based on the fact that at any given time, by controlling the pressures in the hydraulic cylinders cavity, it is possible to achieve a balance of forces acting on the upper platform, i.e., immobility of the upper platform, while the bottom platform can perform unwanted movement. The required pressure in the head-end cavities of the hydraulic cylinders is maintained via injection (normally open) and drain (normally closed) controlled valves (Fig. 9.8). The Figure illustrates the diagram of an automatically controlled hydraulic drive of the platform robot mounted between the protected object 1 and the source 2. Adaptive support drive consists of a base 3, the upper platform 4 and

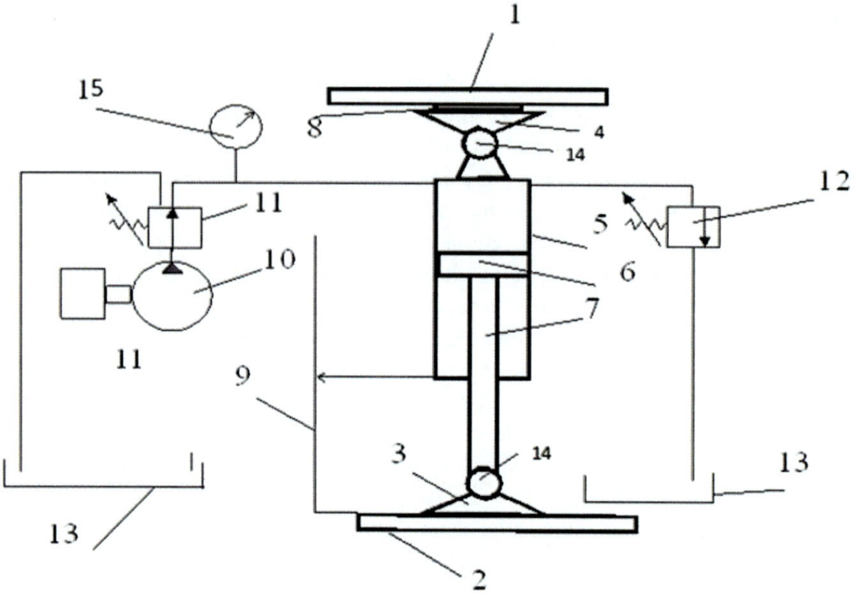

Fig. 9.8 Diagram of the electric-hydraulic drive

the unilateral hydraulic cylinder 5, with the piston 6 and rod 7. Between the upper platform 4 and the object 8, the strain gauge is mounted, which measures the force effecting the platform (Fig. 9.9).

The rod displacement is measured by the displacement sensor 9. The hydraulic system consists of the pump station 10, the normally open pressure control solenoid valve 11 and the normally closed drain solenoid valve 12, the reservoir 13. The pressure in the cylinder cavity is defined by the sensor 15, the upper and bottom platforms are connected to the hydraulic cylinder by spherical joints 14. The hydraulic drive work is based automatic provision of the required pressure in the head-end of the hydraulic cylinder. Depending on the external load determined by the strain gauge 8 and the position of the upper platform determined by the displacement sensor 9, the required fluid pressure in the head-end is calculated according to the algorithm; this pressure is compared to the actual pressure detected by the sensor 15. The input signal to trigger the system is the error signal coming from the sensor 15. With the drain 11 and discharge 12 valves, the desired pressure is provided in the cavity. Figure 9.8 shows the experimental unit of the platform robot, the executive mechanism of which is SHOLKOR MPM. The robot is equipped with pressure sensors 3, magnetostrictive displacement sensors 2, strain gauges 1. The executive devices of the system are injection 4 and drain 5 solenoid valves. The hydraulic system is fed by the hydraulic pump 6. To automate the operation of the robot, controlled hydraulic system is designed as an active support; its functional diagram is shown in Fig. 9.10. The Figure uses the following symbols: IC—industrial controller; DS—displacement sensor; SG—strain gauge; PS—pressure sensor; NO—injection valve

Fig. 9.9 Experimental unit of the platform robot

(normally open), NC—normally closed drain valve. The system operates as follows: the signal from the DS and SG in real-time mode arrives to the IC. IC according to the existing algorithm determines the required pressure in the cylinder cavities RN is compared by the IC to the actual pressure RD according to the data from PS, and the binary signal is supplied to one of the valves, if RD > RN, then the drain valve NC is opened, and the injection valve is closed; otherwise, the injection valve NO is opened, and the drain valve is closed. When the pressures are equal, the valves remain in their position. In the case when the rod is in the uppermost position or the lowermost position, valve state can be changed only in a certain way, or not changed. For example, the rod has reached the uppermost position with RD > RN, it is useless to open the drain valve, hence the state of the valve must not be changed. On the contrary, with RD < RN, it is necessary to enable the injection valve, but to exclude

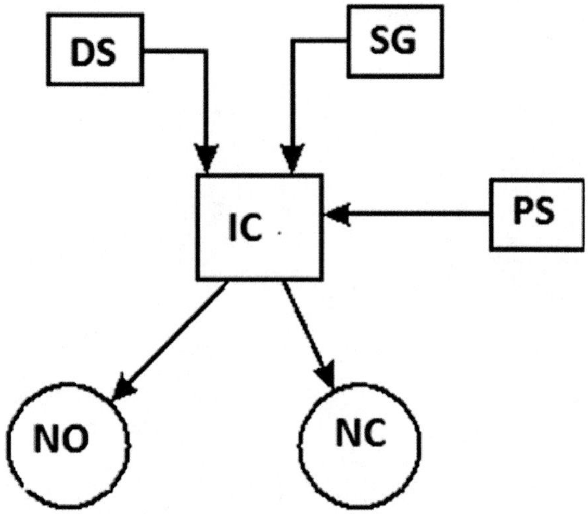

Fig. 9.10 Functional diagram of the controlled hydraulic system

the situation when the pressure RD exceeds the maximum value set by the capacity of the pump station.

In this connection, the signals of the displacement of the corresponding extreme positions must additionally enter the IC to made decisions on the other algorithm. The current system parameters and modes of valves operation of valves are displayed by SCADA visualization system on the computer screen.

Currently, one of the urgent problems of the modern robotics is to create robots, capabilities of which exceed the human potential, which can perform specified functions in extreme conditions or in conditions where human functions are limited. For example, it was found that human possibility associated with the motor reaction (human reaction time) is greater than 200 ms. For some robots, provision of the required motor reaction is the major task in terms of their functionality. For example, if the platform robot is used as an active support, it must quickly respond to the actions of the source of emergency impacts and take actions so that these impacts do not affect the protected object. In this case, the functionality of the adaptive support is largely determined by its motor reaction.

It is known that the motor reaction (MR) is understood as the time of the quickest response by the most simple and predetermined motion to the sudden, but known signal. Hereinafter, it is assumed that the motor reaction of the platform robot is determined by the motor reaction of the controlled hydraulic drives. In this regard, below the question of determining the motor reaction of a single controlled drive is considered first. The controlled hydraulic drive has elements that are different in their physical nature and functions performed. For example: software and hardware means for automatic control; solenoid valve, line bearing the working liquid (WL); hydraulic engine in the form of unilateral action of the hydraulic cylinder with a rod and return spring. These elements jointly effect the motor reaction of the hydraulic drive. In this case, due to the significant highspeed response of the modern microprocessor equipment and, as a result of the small motor reaction of the hardware and software controls, the main focus is given to studying two elements, namely, the solenoid valve, and the power element—hydraulic cylinder. And here it assumed that the motor reaction of these elements depends on the time constant in the case where each of the said elements is regarded as a dynamic link of the control system.

In this case, the motor reaction of the controlled hydraulic drive equals to some fraction of the sum of the time constants of two links: the solenoid valve and the hydraulic engine.To analyze the motor reaction of actual drives, an experimental study of the time dependence of simultaneous changes in the WL pressure in the pressure line and as a result, displacement of hydraulic drive rods. The SHOLKOR robot (Fig. 9.9) is used as the experimental unit. For the experiments, on the top platform of the robot, a load with the predetermined weight of 15 kg was placed. Displacement sensors were placed at the middle position of the rods, when they are extended by a half of the stroke. Using the strain gauges and pressure sensors, the values of the forces acting on the rod and the pressure in the working cavities of the hydraulic cylinders were determined. These values were processed by the controller and used by the program to control the robot. Then, according to the predetermined sequence, the control unit performed control over the motion of the platform robot. Data obtained from the information and measuring system, were processed by the controller according to the set program, and the dependence diagrams for the rod displacement and pressure in the hydraulic cylinder were displayed on the screen in the SCADA system. Some of the experimental data are shown in Fig. 9.11a, b. The Figure presents diagrams of experimental studies obtained for the two drives of the SHOLKOR platform robot (Fig. 6.1), namely, for the hydraulic drive: 6-*a*); *7-b*). The diagrams obtained in SCAD system show the dependences of the displacement (red line) and pressure (blue line) against the time for the period of 2 min. At the same time, it shows the corresponding phases of opening the injection valve. It should be noted that in the hydraulic drive 6, the piston diameter of the hydraulic cylinder is $d_1 = 16$ mm., and in the hydraulic drive 7 the piston diameter is $d_2 = 20$ mm. While both hydraulic drives are unilateral with return spring. In the hydraulic drive 7, the piston has a throttling opening, and the hydraulic drive 6, throttling is not provided. To scale the time in the diagrams obtained in the real time mode, the axis is divided into 620 equidistant segments. Thus, the scale with a division value of 0.19 s is introduced. As a result, these graphs in Fig. 9.11 demonstrate that the movement of the rod begins in 0.76 s after the occurrence of the WL pressure in the cavity of the hydraulic cylinder. When performing the experiments with hydraulic drive 6, the state of the injection valve was further observed. Figure 9.11b shows the time of opening the injection valve, which corresponds to the end of the solid red line. The diagrams show that in 4.5 s after the valve is opened, there is a pressure change in the cavity of the hydraulic cylinders. It is obvious that this delay is due to the resistance in the WL line. By comparing the pressure and displacement diagrams, it is obtained that the rod movement begins in 4.56 s after the occurrence of the pressure in the cavity of the hydraulic cylinder. The resulting time parameters that characterize response motor reactions of different dynamic links comprising the hydraulic drives, allows for the following conclusions: the longest time of the motor reaction belong to the WL line;

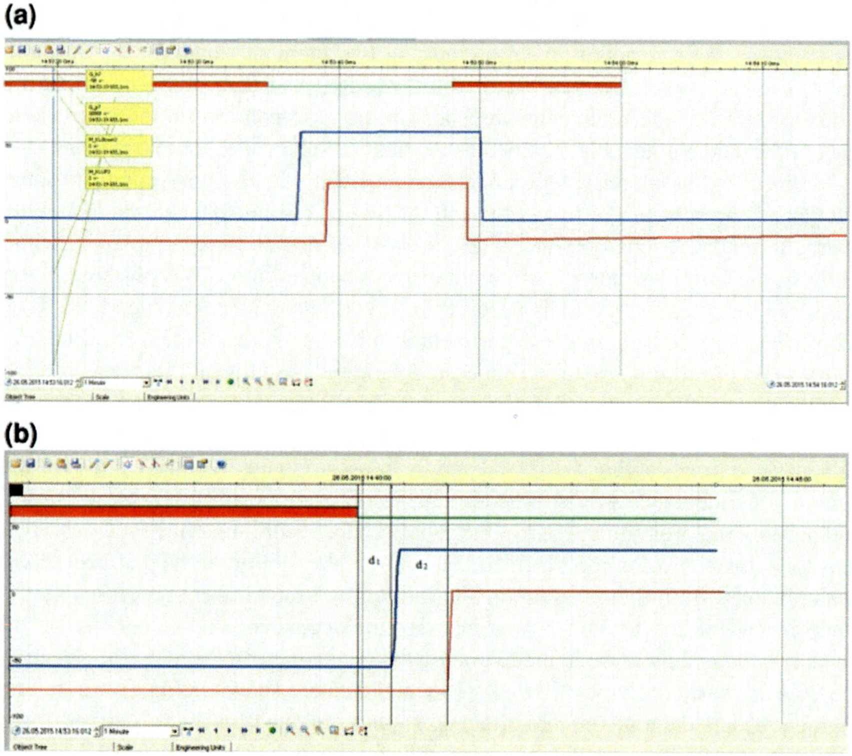

Fig. 9.11 Diagrams of the experimental research

time of the hydraulic drive motor reaction with throttle openings in the piston are 6 times greater than that in the hydraulic drive without throttling. It follows that, improvement of the motor reaction of the drive requires joint performance of the executive devices and control apparatus in the hydraulic drive, i.e., it is necessary to use mechatronic hydraulic drives to exclude the influence of the WL line. On the other hand, the experiments have shown that it is undesirable to use throttling elements in the hydraulic drives of the platform robots used as active supports.

The results of experimental studies have shown that itb is possible to achieve the desired values of motor reaction so that the platform robot can be used as an active support.

9.2.1 Application of the Active Supports Composed of SHOLKOR Robots

The use of the active system of seismic protection of buildings and structures is the most effective in the seismic protection technology. Active seismic protection systems consist of the active controlled supports. Thus, the platform robot consisting of SHOLKOR MPM, discussed above, may be employed in the system of active seismic protection. As an example for a possible application of the platform robot, Fig. 9.12 shows the combined seismic protection system [77] consisting of active and passive seismic protection systems. In this system, between the building superstructure 1 and the foundation 2 with the bottom part of the base 3, passive seismic isolation device, such as kinematic foundation 4 (spheres may be used) is installed. Furthermore, at the lateral sides of the lower part of the building, to settle the horizontal forces of the earth crust action (generally leading to destruction), a system of active seismic protection is used; this system is formed by the SHOLKOR platform robots 5, the number of which is found by calculation. In this system, part of the robot platforms is connected to the foundation 2; during an earthquake, these platforms perform spatial motions due to earth crustal movement. Another part of the robot platforms is connected to the building. Application of the SHOLKOR platforms is based on the fact that in any position of the connecting links and the bottom platform, using the automatically controlled system described above, it is possible to set the

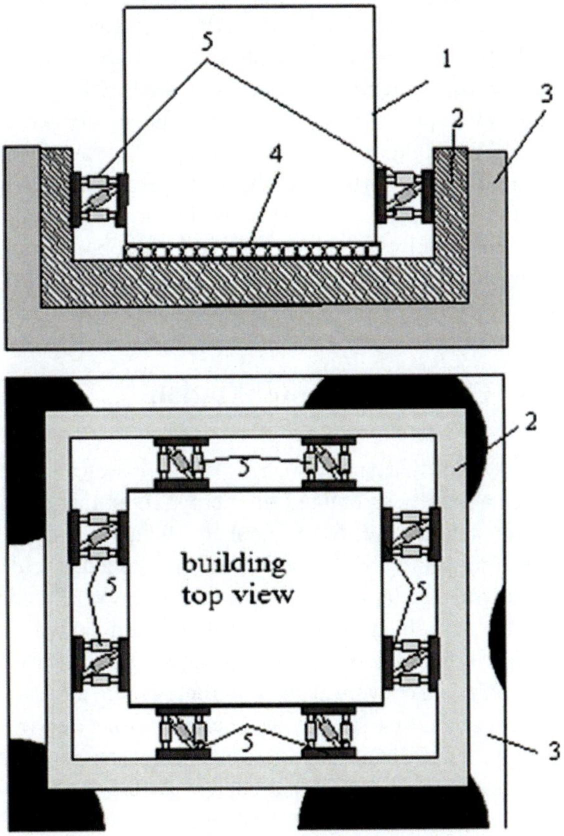

Fig. 9.12 Diagram of the combined seismic protection system

working liquid pressure in the drives cylinders, which balances the loads effecting the upper platform. As a result, at any position of the bottom platform, it is possible to arrange that the upper platform remains in the fixed state. For example, when using the combined system formed of active and passive seismic isolation systems (Fig. 9.12). The base of the building, which in our case is located on the spheres, will move in the case, when the building in the horizontal plane will be effected by the external forces Q exceeding the friction forces $F = f \cdot G$, where f is friction coefficient; G is building weight.

Based on this, efforts exerted by the platform robots together, must not exceed the force F, which is provided by the controlled system of robots. Under the action of the forces in the horizontal area, in case of earthquakes, the bottom platforms of the robot will move along with the foundation, while the robots allow for movement with an amplitude of 1 m. Additionally, the controlled system will exclude the occurrence of resonant oscillations.

In the case of using the combined system composed of active seismic isolation devices (Fig. 9.13) in the static position, the building 1 is supported by the rigid links 7.

If the earthquake signs occur, the robot drives are switched on, which lift the building so that the rigid supports 7 are freed and folded. Subsequently, seismic protection functions are transferred to the controlled bottom platform robots 6, the number of which is selected based on the weight of the building. Moreover, the upper and bottom platforms of these robots are rigidly connected to respectively the lower slab of the building 1 and the bottom of the foundation 2. In this combined system, the platform robots 6, due to the above properties, can maintain balance of their upper platforms 9, i.e., balance of the building, while the bottom platforms will damp the movement of the foundation 2. Platform robots 5 positioned on the horizontal axis compensate the displacements of the foundation 2 in the horizontal plane.

9.3 Float Wave Power Station

It is known that surface waves of large water spaces (oceans, seas, large lakes) have an enormous amount of energy. To extract and use this energy, various wave energy sources (WES) are applied. A fairly complete overview of the design and operating principles of the WES is presented in the paper [78–81]. However, the successful development of wave energy and WES widespread use are prevented by a number of problems, the significant ones among which are their low productivity and susceptibility to destruction by the dynamics of waves and corrosion. In most WES, the process of converting the wave energy into the electrical energy takes place in two stages: at the first stage wave energy is converted into the "organized" mechanical energy; at the second stage, this mechanical energy is converted by a variety of known processes into the electrical energy. At the same time the efficiency of extracting energy of wave motion at the first step depends largely on the functionality of the devices that hereinafter are refereed to as the primary transducers. Most of the existing

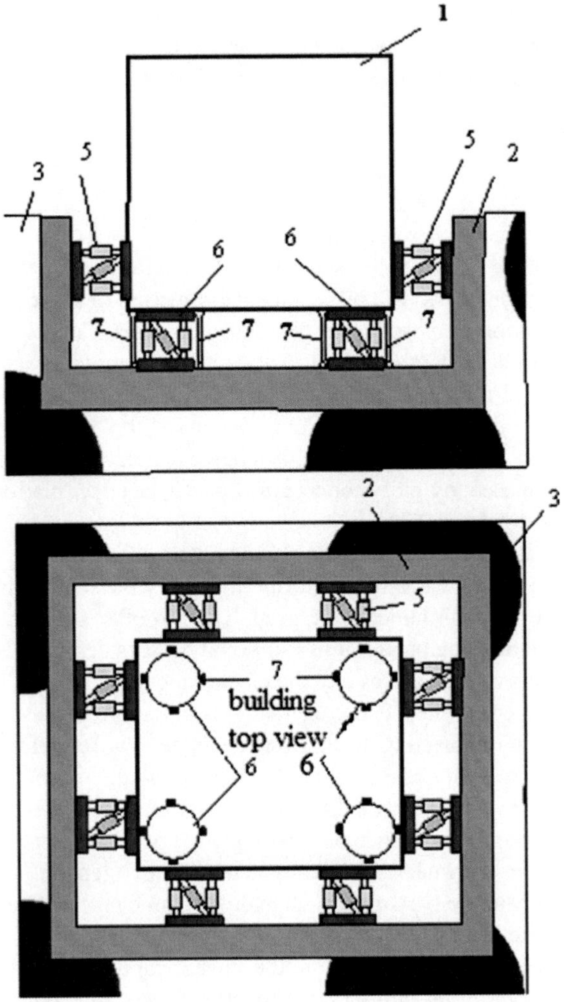

Fig. 9.13 Active seismic protection system

primary transducers have low efficiency coefficient and are not effective, since they only use the energy of the vertical movement of the water masses. In the existing float wave electric power stations (FWEPS) use converters that monitor the wave profile for converting the wave energy. And operation of the transducers that monitor the wave profile, is based on floats of different configuration, which are connected through the rod and the cylinder with the bottom or other floats, and result in the translational motion of the rod when the distance from the point of the float attachment to the bottom of the other float. Thus, according to the existing conversion technology, the movement of the water mass only in vertical direction is converted into a single translational motion. However, in case of commotion, water particles in large water bodies are moving along the circular orbit, parameters of which are determined according to the trochoidal wave theory [82]. It is known that the orbit radius at the lower boundary of the water surface is equal to $r_0 = H/2$, where H is wave height. With the change of the distance (h) from the surface downward, the orbit radius of water particles movement (r) decreases exponentially.

$$r = r_0 e^{-2\pi h/\lambda}. \tag{9.1}$$

where λ is wave length.

The plane of a closed orbit is located vertically to the water surface and perpendicular to a wave crest. Since the direction of wave advance can change over time, position of closed orbit plane also changes. Consequently, the movement of water particles in fact are spatial. The captured by moving water mass float of the submerged FWEPS actually moves along the spacial closed path, performing spatial motions with six degrees of freedom. To convert the energy of all the spatial motions of the float into the "organized" mechanical energy of the six translational motions, it is proposed to rigidly connect the float to the upper platform of the SHOLKOR parallel manipulator [83].

The peculiarity of the selected manipulator converter is that in order to connect the drives with the platforms [32], flexible connections adapted for operation in hostile environments are used. The FWEPS includes the working body (WB) and parallel manipulator and additional devices designed for converting the mechanical energy of the drives into the electrical energy, as well as systems for generating and converting the electric energy. WB in its turn consists of the float and the structure (mast) connecting the float and the upper platform of the manipulator. Feasibility and functionality of the FWEPS using technology of more efficient conversion of wave energy, are confirmed by demonstration model of the FWEPS (Fig. 9.14). Shown here are: 1-bottom manipulator platform; 2-six actuators representing the connection "stem—cylinder"; 3-linear electric current generators, rotors of which move along with the stems; 4-float with hydrodynamic profile; 5-unit ampere meter; 6 aquarium for testing.

Under the wave action, the float is captured by moving water and drives (stem-cylinder connections) are brought into translational motion. The rod movement in the demonstration model starts to move the rotor of the linear generator producing the electric current. The tests of the demonstration model of the wave power station were conducted in the aquarium (Fig. 9.15). Where in, the waves were created by hand, wave height was small, but nevertheless one generator triggered. This is explained by the fact that the movement of the water mass in the aquarium is very simple, only the water level changes due to waves. It is characteristic that if the model is rotated about its vertical axis, the other generator is triggered. Another situation was observed during the test on the shore of a small lake. In this case, two generators were periodically triggered: during the clash of small waves and their pullback. The tests confirmed the functionality of the FWEPS. To arrange protection against the damaging effects of the wave dynamics, it is proposed to install the FWEPS in the coastal area under water so that the float is positioned under the water at a short distance from the surface. In case the location the FWEPS installation is remote

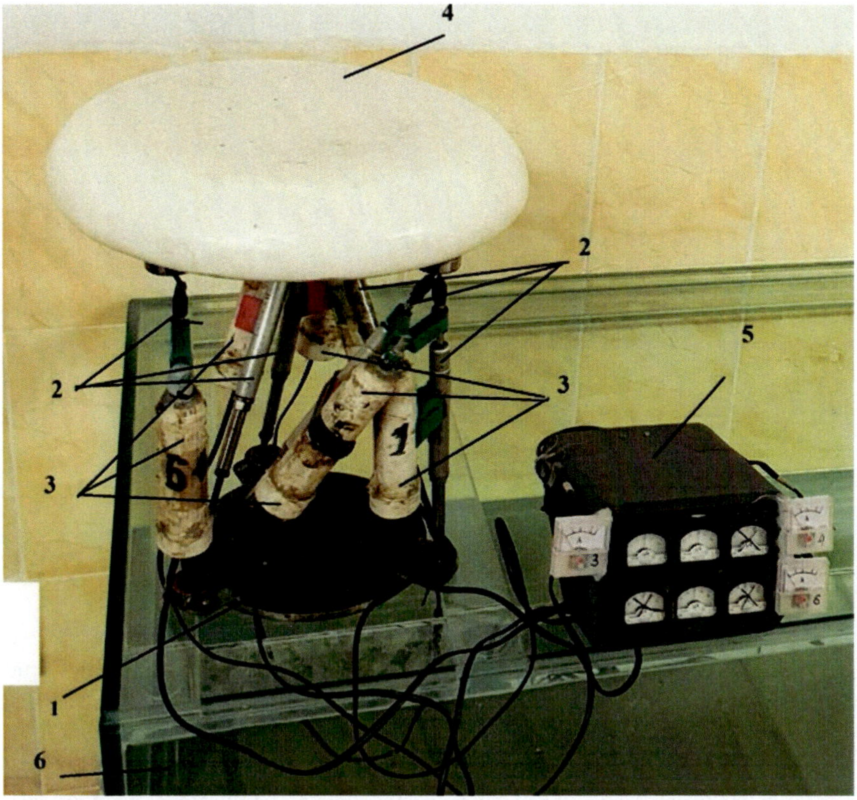

Fig. 9.14 Demonstrational model of the FWEPS

from the coastline, as well as to choose the right position of the float, the mast or frame construction is used, which is rigidly connecting the float and the upper platform of the manipulator. The submerged position of the FWEPS decreases the impact of the wave dynamics having the greatest destructive power at the surface and simultaneously provides allowable spatial motion of the float along with the water mass.

9.4 Sail Wind Power Station Based on the MPM

Today, in the renewable energy sector the direction associated with the use of wind energy develops intensively. Applied there turbine wind power stations (WPS) in terms of their design are divided into those with horizontal and vertical orientation of turbine axes [84]. Moreover, the most popular ones are WPS with the horizontal orientation of the turbine axes, the rotor blades of these turbines are moving under

Fig. 9.15 Demonstration model of the FWEPS during the tests in the aquarium

the effect of the wind lift forces [85]. The wind energy efficiency does not exceed the Betz coefficient equal to 0.593. The operation that does not depend on the wind direction distinguishes the turbine WPS with vertical orientation of the turbine axes. In these WPS turbines work due to the forces of resistance to the air flow [86]. However, the main problem of the turbine WPS of both types is the unpredictability of wind, wind gusts speed and force, which often change within short time intervals. Furthermore, the lower limit of the wind speed range, with which the output power of turbine WPS, usually amounts to 30% of the rated speed, is approximately 8 m/s. While areas with wind speeds ranging from 3 m/s mostly occupy a vast territory with population and industrial facilities. It is in these areas, where the use of turbine WPS appears to be ineffective.

To increase the wind energy efficiency, instead of rotating turbine blades it is proposed to use sail WPS, for example, with a rectangular design of the sail and the vertical rotation shaft [87]. This design is similar to the WPS with the vertical turbine axis. Other sail WPS has a fundamental difference, as it has no rotating turbines [88]. However, in this WPS, all moving devices that support the operation of the WPS, are located on the mast. This creates undesirable dynamic loads that cause the mast oscillation and additional load on the devices and their connections. In order to increase efficiency of the WPS, the new sail WPS [32] is proposed, wherein the working body is a sail with toroidal shape (Fig. 9.16), which swings under the action of the lifting force and resistance force of the wind. The WPS peculiarity consists in that the oscillating motion of the sail is transmitted to the moving platform of the SHOLKOR MPM, which converts these motions into six translational motions. Then the energy of the translational motion by known methods, e.g., by mechanical, pneumatic, hydraulic devices, is converted into the electrical energy. Due to the more

Fig. 9.16 Sail of the WPS

complete utilization of the wind energy, WPS efficiency increases, on the other hand (as compared to the turbine WPS), all massive moving devices are arranged on the fixed base, which increases the reliability and durability of the WPS.

In the proposed sail WPS, the toroidal sail has an aerodynamic profile of its sections, and may be performed in the form of a hollow body made of solid elastic material (parachute fabric) filled with gas or it may be made of lightweight synthetic or composite material. For example, Fig. 9.16 presents a sail made of Penoplex.

Depending on the wind speed, the volume of the hollow sale changes by pumping or discharging the gas or by other means allowing to change the surface exposed to the wind. This kind of the sail allows WPS to start functioning at a wind speed of 1–1.5 m/s. Under the action of the wind, the sail performs spatial periodic motions depending on the wind direction and speed.

In the WPS the sail is located at height of more than 10 m and is rigidly connected to the upper platform of the SHOLKOR MPM [30]. *This manipulator allows converting the spatial motion of the sail into six translational motions.* Fig. 9.17 shows the demonstration model of the sail WPS.

Here the fixed platform 1 of the parallel manipulator is connected to the upper platform 2 by six drives 3. Each rod of the rod-cylinder connection of actuators 3 is linked to the linear power generator rotor 4 that generates electric current registered by the microammeters block 7. The peculiarity of the MPM model consists in that the drives with platforms are fixed by flexible connections protected by the patent [32], and allowing to operate the complex connections without lubrication in an open air environment. In the model, the sail is located at a small height, in addition, the sail has aerodynamic profile in cross section and did not change its surface exposed to the wind. However, this model allowed to demonstrate the principle of WPS operation. In detail, under the action of the wind, the sail performed spatial periodic motions depending on the wind direction and speed. Thus, the sail captures the kinetic energy of the air mass, and the manipulator converter of the MPM converts this energy into mechanical energy of translational motions of the six cylinder rods, which allows

Fig. 9.17 Operational model of the WPS

generating the electric energy. On the model, it can be seen that the sail moving under the wind force, at the same time compresses the spring in the rod-cylinder connection. During the return stroke, the sail moves under the effect of the spring. Also the operational model showed that the shape of the sail in the form of a torus prevents WB from the destructive action of the wind. For example, in case of a strong winds, WB deflected for the maximum angle, at which stall of the airflow occurred due to the toroidal shape of the sail. Once the wind gust disappeared, the WB returned to its original position under the action of elastic forces. 2–3 telescopic connections actively operated with the given wind direction. When the wind direction changed, other 2–3 telescopic connections functioned. The movement of the working body in the absence of strong winds was represented by oscillations with a small amplitude. The essential difference of the prototype from the existing WPS is the use of the swinging sail with automatically controlled surface exposed to the wind and MPM with six degrees of freedom.

Conclusion

The analysis of the kinematic chains of the manipulators with open kinematic chains isolated the core functional groups (CFG). Various interconnections of the CFG provided the architecture of one-loop parallel manipulators (OPM). Application of the methods of the combinatorial topology to the analysis of the relative motions of the connections of two links allowed creating the topological model of the kinematic pair. Based on the topological model of the six-DOF kinematic pair, the architecture of the SHOLKOR multiloop parallel manipulator (MPM) was created. For the OPN and MPM, design, kinematics and dynamics analysis methods were provided. At the same time, to analyze the design and kinematics the paper proposes original universal methods.

The study also proposes the possible fields of application of OPM and MPM, which are confirmed using the operational demonstration models.

K. S. Sholanov, *Parallel Manipulators of Robots*, Mechanisms and Machine Science 92,
https://doi.org/10.1007/978-3-030-56073-7

References

1. Angeles, J.: Fundamentals of robotic mechanical systems: theory, methods and algorithms, 2nd edn, p. 520. Springer, (2002)
2. Merlet, J.P.: Parallel Robots. Springer Publishers, Dordrecht (2006)
3. Ceccarelli, M.: Fundamentals of the Mechanics of Robots. Springer, (2004)
4. Angeles, J.: The qualitative synthesis of parallel manipulators. ASME J. Mech. Des. **126**(4), 617–624, Juillet (2004)
5. Alizade, R., Selvi, O., Gezgin, E.: Structural design of parallel maniulators with general constraint one. Mech. Mach. Theory. **45**(1), 1–14, Janvier (2010)
6. Balchanowski, J.: General method of structural synthesis of parallel mechanisms. Arch. Civ. Mech. Eng. **10** (2016)
7. Rico, J. M. et al.: New considerations on the theory of type synthesis of fully parallel platforms. ASME J. Mech. Des. **130**(11), 112302–1/9, (Nov 2008)
8. Simoni, R., Piga Carboni, A., Martins, D.: Enumeration of parallel manipulators. Robotica. **27**(4), 589–597, Juillet 2009
9. Fan, C., Liu, H., Zhang, Y.: Type synthesis of 2T2R, 1T2R and 2R paralllel manipulators. Mech. Mach. Theory. **61**, 184–190, Avril 2013
10. Fang, Y., Tsai, L.-W.: Structure synthesis of a class of 3-DOF rotational parallel manipulators. IEEE Trans. Robot. Autom. **20**(1), 117–121, February 2004
11. Gogu, G.: Structural Synthesis of Parallel Robots, Part 3: Topologies with Planar Motion of the Moving Platform. Kluwer, Dordrecht (2010)
12. Gogu, G.: Structural Synthesis of Parallel Robots, Part 4: Other Topologies with two and Three Degree of Freedom. Sringer, Dordrecht (2012)
13. Sholanov, K. S.: To analyze the structure of kinematic chains of manipulators. Robots and Robotic Systems, Proceedings of the Bauman Moscow State Technical University, Moscow, 1983
14. Sholanov, K. S.: On a method for synthesizing kinematic chains of manipulators using SFG. Robots and Robotic Systems, Proceedings of the Bauman Moscow State Technical University, Moscow, 1985
15. Baigunchekov, J., Sholanov, K.: Structural and parametric synthesis of new parallel manipulators with functionally independent drives. International Con-ference on Mechatronics, Kuala Rumpur, Malaysia (2001)
16. Wang, X., Baron, L., Cloutier, G.: Topology of serial and parallel manipulators and topological diagrams. Mech. Mach. Theory. **43**(6), 754–770, Juin 2008
17. Sholanov, K. S.: A manipulator. Author certificate USSR №1289675, Moskow, 1986

K. S. Sholanov, *Parallel Manipulators of Robots*, Mechanisms and Machine Science 92, https://doi.org/10.1007/978-3-030-56073-7

18. Sholanov, K. S.: Manipulator of a six-coordinate table. Author certificate USSR. № 93171, Moscow, 1993
19. Sholanov, K. S. The Walking Apparatus. Author certificate USSR №1716702, Moskow, 1989
20. Jin, Q., Yang, T.-L.: Theory for topology synthesis of parallel manipulators and its application to three-dimension-translation parallel manipulators. ASME J. Mech. Design. **126**(1), 625–639, Janvier 2004
21. Hess-Coelho, T. A.: Topological synthesis of a parallel wrist manipulator. ASME J. Mech. Des. **128**(1), 230–235, Janvier 2006
22. Lu, N., Ye, Y.: Type synthesis of parallel mechanism by using sub-mechanisms and digital topological graphs. Mech. Mach. Theory. **109**, 39–50 (2017)
23. Spanier, E.: Algebraic topology. p. 680, Moscow, Mir, 1971
24. Pontryagin, L.S.: Fundamentals of Combinatorial Topology, p. 136. Nauka, Moscow (1976)
25. Sholanov, K.S.: Simplicial Method for Analyzing and Synthesizing the Structure of Mechanisms, p. 86. Gylym, Amaty (2000)
26. Sholanov, K. S.: Application of topological model for analysis of structure of mechanisms and manipulators with parallel structure. The 2nd IFToMM Asian Conference on Mechanism and Machine Science November 7 -10, Tokyo, Japan, (2012)
27. Wittenburg, J.: Dynamics of systems of solid bodies. M: Mir, 1980
28. Malyshev. A. P.: Analysis and synthesis of mechanisms from the point of view of their structure. Bulletin of the Tomsk Technological Institute, vol. 44, p. 78. Tomsk (1923)
29. Denavit, J., Hartenberg, R. S.: A kinematic notation for lower pair mechanisms based on matrices. J. Appl. Mech. **22**, Trans. ASME, Vol. 77, Series E, pp. 215–221
30. Sholanov, K.: Platform robot manipulator. WO/2015/016692. 05 Feb 2015
31. Sholanov, K. S.: Multi-Link spherical connection (variants). Innovation patent no. 27181, bul. no. 7,15. 07 (2013)
32. Sholanov K. S. Power plants (variations) on the basis of parallel manipulator. WO/2018/147716, 16.08.2018
33. Gogu, G.: Mobility of mechanisms: a critical review. Mech. Mach. Theory. **40**(9), 1068–1087 (2005)
34. Voinea, R., Atanasiu, M.: Contribution a l' etude de la structure des chanies cinematiques. Bull. Ynst. Politehn. Bucuresti, t.XXII, Fasc.1, 1960
35. Angeles, J.: The degree of freedom of parallel robots: a group-theoretic approach. In IEEE Int Conf on Robotics and Automation, pp. 1017–1024, Barcelona, 19–22 Avril 2005
36. Lu, N., Ye, Y.: Type synthesis of parallel mechanism by using sub-mechanisms and digital topological graphs. Mech. Mach. Theory. **109**, 39–50 (2017)
37. Frisoli, A. et al.: Synthesis by screw algebra of translating in-parallel actuated mechanisms, In ARK, Piran, pp. 25–29 Juin 2000
38. Craig, J. J.: Introduction to Robotics. Mech. Control. Pearson Education International (2013)
39. Yang, T.-L., Sun, D.-J.: A general DOF formula for parallel mechanisms and multi-loop spatial mechanisms. ASME J. Mech. Robot. **4**, 011001-1-17 (2012)
40. Yang, T.-L., Sun, D.-J.: A general DOF formula for parallel mechanisms and multi-loop spatial mechanisms. ASME J. Mech. Robot. **4**, 011001-1-17 (2012)
41. Rocha, C.R., Tonetto, C.P., Dias, A.: A comparison between the Denavit-Hartenberg and the screw-based methods used in kinematic modeling of robot manipulators. J. Robot. Comput. Integr. Manuf. **27**(4), 723–728 (2011)
42. Sheth, P. N., Uicker J. J. Jr.: A generalized symbolic notation for mechanisms. J. Eng. Indus. Trans. ASME, Series B, **93**(1), (Feb 1971)
43. Sholanov, K. S.: Multi-moving mechanisms with closed kinematic chains, p. 149, Almaty. Gylym, 1999
44. Dai, Jan S.: An historical review of the theoretical development of rigid body displacements from Rodrigues parameters to the finire twist. Mech. Mach. Theory. **41**(1), 41–52 (2006)
45. Sholanov, K. S.: Eight parameters method for kinematic modeling of robots' actuating mechanisms. Proceeding of 11th World Congress in Mechanism and Machine Science, vol. 1–5, pp. 121–123. Tianjin, China (2004)

46. Kane, T. R., Levinson, D. A.: The use of Kant's dynamical equations in robots. Intern. Journ. Robot. Res. **2**(2), 3–5 (1983)
47. Fu, K., Gonzalez, R., Lee, C.S.G.: Robotics: Control, Sensing, Vision and Intelligence. McGraw-Hill, New York (1987)
48. Lling, S., Huang, M.: Kinestatic analysis of general parallel manipulators. J. Mech. Des. **117**(4), 601–606 (1995)
49. Korendyasev, A. I., Salamandra, B. L., Tives, L. I. et al.: Robot manipulation systems. Under the General Editorship of A. I. Korendyasev, p. 472. Mashinostroenie, Moscow (1989)
50. Sholanov, K.S., Kabanbayev, A., Abzhaparov, K.M.: Study and selection of parameters of automatically controlled wind power station with swaying sails. Int. J. Renew. Energ. Res. **11**(2), 723–737 (2020)
51. Lling, S., Huang, M.: Kinestatic analysis of general parallel manipulators. J. Mech. Des. **117**(4), 601–606 (1995)
52. Xi, F., Zhang, D., Mechefske, C. M., Lang S. Y. T.: Global kine to static modelling of tripod-based parallel kine matic machine. Mech. Mach. Theory, **39**(4), 357–77 (2004)
53. Tsai, L.-W.: Solving the inverse dynamics of a Stwart-Gough manipulator by the principle of virtual work. ASME J. Mech. Des. **122**(1), 3–9 (2000)
54. Li, C.X., Li, C.N.: Dynamic analysis of the GoughStewart platform manipulator. IEEE Trans. Robot. Autom. **16**(1), 94–98 (2000)
55. Pedrammehr, S., Mahboubkhah, M., Pakzad, S.: An improved solution to the inverse dynamics of the general Stewart platform. IEEE Int. Conf. Mech. (2011)
56. Luh, J. Y. S., Walker, M. W., Paul, R. P. C.: On line computational scheme for mechanical manipulators. Trans. ASME. J. Dyn. Syst. Meas. Control, **102**(2), 69–76 (1980)
57. Zhang, C., Song, S.: An efficient method for inverse dynamics of manipulators based on the virtual work principle. J. Robot. Syst. **10**(5), 605–627 (1993)
58. Paul, R.: Modeling, trajectory planning and motion control of the robot manipulator, p. 238. Moscow, Science. (1976)
59. Tardieu, G.: Therapeutic training of psychomotor invalids littles disease ataxia spasticity. Conf. Cours. Med. Vun. **85**(9), 35 (1962)
60. Fregni, F., Imamura, M., Chien, H.F., Lew, H.L., Boggio, P., Kaptchuk, T.J., Riberto, M., Hsing, W.T., Battistella, L.R., Furlan, A.: Challenges and recommendations for placebo controls in randomized trials in physical and rehabilitation medicine: a report of the international placebo symposium working group. Am. J. Phys. Med. Rehabil. **89**(2), 160–72 (2010)
61. Veretelnik, O. V.: Research of various prosthetic structures in prosthetics of the cervical spine.. Visnik NTU "KHPI". TEM. VIP.: "Mashinoznavstvo TA CAD". №12. pp. 18–24 (2009)
62. Gough, V. E.: Contribution to discussion of papers on research in automobile stability, control and tyre performance. Proc. Auto Div. Inst. Mech. Eng. pp. 392–394 (1956)
63. Stewart, D.: A platform with six degrees of freedom. Proceedings of the Institution of Mechanical Engineers, vol. 180(15), pp. 371–385. London (1965)
64. Dasgupta, B., Mruthyunjaya, T. S.: The Stewart platform manipulator: a review. Mech. Mach. Theory. pp. 15–40 (2000)
65. Müller, C. W., Otte, D., Decker, S., Stübig, T., Panzica, M., Krettek, C., Brand, S.: Vertebral fractures in motor vehicle accidents–a medical and technical analysis of 33,015 injured front-seat occupants. Accid. Anal. Prev. **66**, 15–19 (May 2014)
66. Servadio, P., Marsili, A., Belfiore, N. P.: Analysis of driving seat vibrations in high forward speed tractors. Biosyst. Eng. **97**(2), pp. 171–180 (June 2007)
67. Shekhovtsov, V. V., Pobedin, A. V., Lyashenko, M. V., Shekhovtsov, K. V.: Classification and analysis of the ways of cabin suspension vibroinsulators' characteristic formation.: XXIX Seminarium KółNaukowych "Mechaników", Warszawa,/WojskowaAkademiaTechniczna. pp. 449–452, Warszawa (2010)
68. Aggromito, D., Chen, B., Thomson, R., Wang, J., Yan, W.: Effects of body-borne equipment on occupant forces during a simulated helicopter crash. Int. J. Ind. Ergonomics. **44**(4), pp. 561–569 (July 2014)

69. Kim, K. S., Kim, J., Kim, K. J.: Dynamic modeling of seated human body based on measurements of apparent inertia matrix for fore-and-aft/vertical/pitch motion. J. Sound Vib. **330**(23), pp. 5716–5735 (Nov 2011)
70. Cation, S., Jack, R., Oliver, M., Dickey, J. P., Lee-Shee, N. K.: Int. J. Ind. Ergonomics. **38**(9–10), pp. 739–757 (Sept–Oct 2008)
71. Wang, A. P., Lin, Y. H.: Vibration control of a tall building subjected to earthquake excitation. J. Sound Vib. **299**(4–5), 757–773 (Feb 2007)
72. Pozo, F., Zapateiro, M., Acho, L., Vidal, Y., Luo, N.: Experimental study of semiactive VSC techniques for vehicle vibration reduction by. J. Franklin Inst. **350**(1), pp. 1–18 (Feb 2013)
73. Uys, P.E., Els, P.S., Thoresson, M.: Suspension settings for optimal ride comfort of off-road vehicles travelling on roads with different roughness and speeds. J. Terramech. **44**(2), 163–175 (Apr 2007)
74. Danh, L. T., Ahn, K. K.: Active pneumatic vibration isolation system using negative stiffness structures for a vehicle seat. J. Sound. Vibr. **333**(5), 1245–1268 (28 Feb 2014)
75. Filipović, D., Schröder, D.: Control of vibrations in multi-mass systems with locally controlled absorbers. Automatica. **37**(2), 213–220 (Feb 2001)
76. Aggromito, D., Chen, B., Thomson, R., Wang, J., Yan, W.: Effects of body-borne equipment on occupant forces during a simulated helicopter crash. Int. J. Ind. Ergonomics. **44**(4), 561–569 (July 2014)
77. Sholanov, K. S.: Combined earthquake protection system (variants). WO2015/099519A1, 02.07.2015
78. Wave energy technology brief. IRENA Ocean Energy Technology Brief 4, June 2014
79. Mei, C. C., Stiassnie, M. & Yue, D. K.-P. Theory and application of ocean surface waves.Singapore: World Scientific, 2005
80. Falnes, J.: Ocean Waves and Oscillating Systems. Cambridge University Press, Cambridge, UK (2002)
81. Cruz, J.: Ocean Wave Energy. Springer, Berlin, Germany (2008)
82. Bezrukov Yu. F. Oceanology. II ii. dynamic phenomena and processes in the ocean. Simferopol: V. I. Vernadsky Tauride national University, p. 123 (2006)
83. Sholanov, K. S.: Wave electric power station on the basis of a parallel manipulator. WO/2017/003273, 05.01.2017
84. P'yankov, K. S., Toporkov, M. N.: Mathematical modeling of flows in wind turbines with a vertical axis. Fluid Dyn. **49**(2), 249–258. SS (2014)
85. Gasch, R., Twele, J. (Hrsg.) Windkraft Langer Grundlagen, Entwurf, Planung und Betrieb, Springer, **XXIII**, p. 599 (2016)
86. Apelfröjd, S., Eriksson, S., Bernhoff, H.: A review of research on large scale modern vertical axis wind turbines at uppsala university. Energies. **9**(7), 570 S (2016)
87. US20140341736A1, Nov. 20, 2014, Sail wind turbine
88. US 2013 0181458A1, Jul.18, 2013, System for converting wind energy